W0106168

Lecture Notes in Economics and Mathematical Systems

For information about Vols. 1–156, please contact your bookseller or Springer-Verlag

continuation on page 139

Lecture Notes
in Economics and
Mathematical Systems

Managing Editors: M. Beckmann and W. Krelle

307

Theo K. Dijkstra (Ed.)

On Model Uncertainty and its
Statistical Implications

Proceedings of a Workshop, Held in Groningen,
The Netherlands, September 25–26, 1986

Springer-Verlag

Berlin Heidelberg New York London Paris Tokyo

Editorial Board

H. Albach M. Beckmann (Managing Editor) P. Dhrymes
G. Fandel G. Feichtinger J. Green W. Hildenbrand W. Krelle (Managing Editor)
H. P. Künzi K. Ritter R. Sato U. Schittko P. Schönfeld R. Selten

Managing Editors

Prof. Dr. M. Beckmann
Brown University
Providence, RI 02912, USA

Prof. Dr. W. Krelle
Institut für Gesellschafts- und Wirtschaftswissenschaften
der Universität Bonn
Adenauerallee 24–42, D-5300 Bonn, FRG

Editor

Dr. Theo K. Dijkstra
University of Groningen, Institute of Econometrics
P.O. Box 800, 9700 AV Groningen, The Netherlands

ISBN-13: 978-3-540-19367-8   e-ISBN-13: 978-3-642-61564-1
DOI: 10.1007/978-3-642-61564-1

This work is subject to copyright. All rights are reserved, whether the whole or part of the material
is concerned, specifically the rights of translation, reprinting, re-use of illustrations, recitation,
broadcasting, reproduction on microfilms or in other ways, and storage in data banks. Duplication
of this publication or parts thereof is only permitted under the provisions of the German Copyright
Law of September 9, 1965, in its version of June 24, 1985, and a copyright fee must always be
paid. Violations fall under the prosecution act of the German Copyright Law.

© Springer-Verlag Berlin Heidelberg 1988

Editor's introduction

The assessment of the quality of models, estimators or predictors in applications is a subtle and complex task. It is in general a lot more difficult than standard textbooks tend to suggest. One of the main sources contributing to its complexity springs from the fact that typically the data at hand are used to select the model. This violates what Leamer (1979) calls the axiom of specification. And this is something no existing statistical school, be it bayesian or frequentist in spirit, can easily handle, see e.g. Leamer (1979), Mosteller & Tukey (1977), Dawid & Dickey (1977). From a frequentist perspective (all contributions to this volume are written in a frequentist mood), the problem is simply this: in deriving frequency distributions of statistics we cannot assume that the same definitions will be used for every point in the sample space, since different samples may favour different models. In general, this makes the derivation of frequency distributions a very complicated task. In fact it cannot even in principle be accomplished unless the model selection rule used induces a partitioning of a well defined sample space. If this rather restrictive assumption is satisfied several authors have suggested to employ sample re-use methods like the bootstrap to estimate frequency distributions, see e.g. Efron & Gong (1983), Dijkstra (1983), Verbeek (1984) to name just a few. In the context of linear models e.g. this amounts to a simple Monte Carlo study: the relevant frequency distributions are estimated via simulations based on, in general, the empirical distribution function of the estimated residuals of the "largest" model.

Some of the contributions to this volume deal explicitly with the usefulness of the bootstrap in situations where the model is data-driven in a well defined way: Freedman et al., Dijkstra & Veldkamp, and Folmer. *Freedman et al* take as their starting point the linear regression model with a "large" number of potentially explanatory variables, i.e. large relative to the number of observations. Adopting a simple variable selection rule they investigate the induced distortion of a number of conventional goodness-of-fit measures. In addition they study among other things the performance of the bootstrap as a means to eliminate these distortions. Disappointingly, it did not perform very well. A similar result is found by *Dijkstra & Veldkamp* who analyze the sampling distributions of regression coefficients as induced by a popular variable selection rule, for a very small set of explanatory variables. Both papers ascribe the disappointing performance of the bootstrap essentially to the way in which it is implemented, namely via the least squares residuals of the most general model. So any sensible selection rule will be biased in favour of large models in bootstrap replications. Dijkstra & Veldkamp suggest that the bootstrap using the least squares residuals of the true model will perform better, especially when the sample identified all relevant variables. Some Monte Carlo results, not reported here, support this. However, the suggestion has clearly theoretical value only. Freedman et al propose to try to adjust the length of the regression coefficients vector. Their results are not discouraging. Perhaps one could use Stein–rule estimates, a data–based compromise between regression estimates for the "largest" and the "smallest" model, see e.g. Judge & Bock (1978). *Folmer* considers pre–testing with respect to autocorrelation in linear models. He investigates whether the bootstrap can estimate the sampling distribution of the pre–test estimator for the regression and autocorrelation coefficient in a simple model. Here Durbin estimates are calculated in case the hypothesis of no autocorrelation is rejected at conventional critical levels for the Durbin–Watson statistic. Otherwise simple least squares is used. The bootstrap simulations are based on the most general model, i.e. the residuals are determined by the Durbin estimates, always assuming autocorrelation. Again, the bootstrap does not perform very well. This time it may be due to severe underestimation of the autocorrelation coefficient, so that bootstrap samples favour zero autocorrelation disproportionately.

Snijders focusses on another sample re–use method, cross–validation, in the context of time series models. More specifically, he investigates the merits of cross–validation and classical approaches with regard to squared prediction error estimation. He finds that for a certain class of models and predictor families cross–validation does not outperform a more conventional approach with respect to squared prediction error estimation. Moreover cross–validation did not mitigate the effect of chance capitalization as induced by selecting the predictor family that happened to perform best. Snijders also discusses problems inherent to the cross–validatory assessment of cross–validatory choice.

Luijben et al discuss some of the ways in which a covariance structure can be modified, simplified or expanded in the light of the data. With a Monte Carlo study they look into the usefulness of some of the popular model selection statistics, starting from a simple "false" model. In addition they describe the conditional distribution of a certain estimator, conditional upon the event that the selection statistics identify the true model. They find substantial differences between this conditional distribution and the distribution obtained by always using the correct model. This counters the usual retort made by researchers when asked why they ignore the model uncertainly: one always proceeds from the assumption that the model selected is correct. Luijben et al show that one cannot ignore the fact that the sample favoured the true model; see also Dijkstra & Veldkamp.

Steerneman & Rorijs demonstrate the consequences of datamining with repect to forecasting, cf. also Mayer (1975). In addition they give a theoretical discussion concerning the choice of the optimal number of predictors, assuming the predictors are selected sequentially in a fixed order.

De Leeuw attacks the relevance of the concept of "the true model". He would prefer "useful" to "true". A model may be more useful than a competitor in the sense that the estimation or prediction errors are "smaller". And this can be judged by means of estimated average values of suitably chosen measures of distance between data and model space. De Leeuw shows in detail for multinomial experiments that the jackknife can be conveniently used to estimate these averages; there is no need to assume that the model used is "true". Dijkstra (1983) also pointed at the potential usefulness of methods like the jackknife and the bootstrap for the assessment of certain statistics without assuming the model to be correct. De Leeuw does not deal explicitly with chance capitalization. However, the stubborn fact remains that the apparent quality of performance of the apparently best model will tend to be an optimistic estimate of "true" performance. How to correct this is an open question. Unfortunately, the results reported in this volume for sample re–use methods are on the whole not encouraging. One could perhaps say provocatively that model uncertainty cannot be ignored but is impossible to take into account without new data.

Financial support by the Netherlands organization for the advancement of pure research (Z.W.O.) is gratefully acknowledged.

References

Dawid, A.P., and Dickey, J.M. (1977): Likelihood and bayesian inference from selectively reported data. *Journal of the American Statistical Association*, 72, 845–850.

Dijkstra, T.K. (1983): Some comments on maximum likelihood and partial least squares methods, *Journal of Econometrics*, 22, 67–90.

Efron, B. and Gong, G. (1983): A leisurely look at the bootstrap, the jackknife and cross–validation, *The American Statistician*, 37, 36–48.

Judge, G.G. and Bock, M.E. (1978). The statistical implications of pre–test and Stein–rule estimators in econometrics, North–Holland, Amsterdam.

Leamer, E.E. (1979): Specification searches, J. Wiley, N.Y.

Mayer, T. (1975): Selecting economic hypotheses by goodness of fit, *The Economic Journal*, 85, 877–883.

Mosteller, F. and Tukey, J.W. (1977): Data analysis and regression, Addison–Wesley, Massachusetts.

Verbeek, A. (1984): The geometry of model selection in regression. In Dijkstra, T.K., ed.: Misspecification Analysis, Springer, Berlin.

TABLE OF CONTENTS

Editor's introduction

ON THE IMPACT OF VARIABLE SELECTION IN FITTING REGRESSION EQUATIONS

D.A. Freedman, W. Navidi and S.C. Peters
Statistics Department
University of California
Berkeley, Ca 94720

Abstract.

Consider developing a regression model in a context where substantive theory is weak. Search procedures are often used to develop the equation: eg, fitting the equation, dropping insignificant variables, and refitting. As is well known, this can seriously distort the conventional goodness-of-fit statistics. Furthermore, the bootstrap and jackknife may not help in high-dimensional cases.

1. Introduction.

When regression equations are used in empirical work, the ratio of data points to parameters is often low. Further, the exact form of the equation is seldom known *a priori*, so investigators will often do some preliminary screening before settling on the final version of the equation. One stylized version of this strategy is as follows:

(i)   Fit the equation with all variables included.

(ii)   Screen out variables whose coefficients are insignificant at the 25% level. (This level is used to represent "exploratory" analysis.)

(iii)   Refit the equation on the remaining variables.

Real investigators use more complicated – and subjective – screening procedures; the version just presented is mechanical, and therefore amenable to statistical analysis.

As is well known, screening procedures introduce substantial distortion into the conventional measures of goodness-of-fit, like R^2, t or F. See (Lovell, 1983) or (Freedman, 1983), and (Gong, 1986) on logistic regression. Perhaps the bootstrap or jackknife can be used to eliminate these distortions? This question will be investigated here by simulation.

Consider the basic linear model

$$Y = X\beta + \epsilon. \tag{1}$$

Here, X is an $n \times p$ matrix of iid $N(0,1)$ variables; and ϵ is another $n \times 1$ vector of iid $N(0,\sigma^2)$ variables. These distributional facts are known to the investigator. The $n \times 1$ vector Y is computed from (1). The investigator observes X and Y, but not ϵ. The $p \times 1$ vector β of parameters is unknown, as is σ^2, and these are to be estimated from the data.

Two statistical tasks are considered:

(i)   *Estimation.* The object is to estimate β_1; and $\beta_2,...,\beta_p$ are introduced to control other sources of variation and improve the precision in estimating β_1. This is like a standard problem in clinical trials: β_1 is the treatment effect, and columns 2,3,... in X represent covariates.

(ii) *Prediction.* Let ξ be another $1 \times p$ row vector of iid $N(0,1)$ variables, and δ an independent $N(0,\sigma^2)$ variable; δ is unobservable. Suppose

$$\eta = \xi\beta + \delta. \tag{2}$$

The β-vector here is the same as in (1), and is unknown to the statistician. The object is to predict η from ξ, using the β estimated from (1). The explanatory variables ξ are related to the dependent variable and should therefore help in predicting η. This is like a standard problem in econometrics.

Our setup is a statistical cartoon, but it has elements of realism. And in some respects, it provides a favorable environment for conventional methodology. After all, (1) is the textbook regression model: ordinarily, variables will not be normal nor regressions linear. In the simulations, we usually set $\sigma^2=1$, $n=100$ and $p=75$. The number of columns in X may seem large, but in practice an indefinitely large number of covariates present themselves to empirical workers. For example, in typical econometric macro-models, there will be several hundred equations to explain several hundred endogenous variables, but only several dozen data points. The "constraints," including the decision as to which explanatory variables to put in each equation, are largely data-driven. Also see (Freedman, 1981a) or (Freedman-Rothenberg-Sutch, 1983).

We consider β's of the form $\beta_j=\gamma$ for $j=1,...,p_1$ and $\beta_j=0$ otherwise. The γ's of interest are those near the resolving power of the system, ie, of order $\sigma/\sqrt{n-p}$. Indeed, let V_j be the (j,j)-element of $(X^TX)^{-1}$. On our assumptions, V_j is distributed as $1/\chi^2_{n-p+1}$, and so is of order $1/n-p$.

Denote the columns of X by X_j, for $j=1,...,p$. The screening procedure selects a subset S of these columns to enter the equation, as follows:

Fit Y to X by OLS (ordinary least squares), so $\hat{\beta}=(X^TX)^{-1}X^TY$, while $\hat{\epsilon}=Y - X\hat{\beta}$ is the residual vector, and $\hat{\sigma}^2=||\hat{\epsilon}||^2/(n-p)$ is the usual unbiased estimate of σ^2. (3i)

Enter X_1 into the equation automatically. For $j=2,...,p$, enter X_j if $|\hat{\beta}_j|/\hat{\sigma}\sqrt{V_j}$ exceeds the 25%-point of the t-distribution with $n-p$ degrees of freedom: recall that V_j is the (j,j)-element of $(X^TX)^{-1}$. Write $j \in S$ if column j was entered. Then S is a random subset of $\{1,...,p\}$ and $1 \in S$. (3ii)

Let X_S be the matrix consisting of the columns of X which were entered in step (ii). Let p_S be the number of such columns. Refit Y on X_S by OLS, so $\overset{\circ}{\beta}=(X_S^TX_S)^{-1}X_S^TY$. Define $\overset{\circ}{\epsilon}=Y - X_S\overset{\circ}{\beta}$, and $\overset{\circ}{\sigma}^2=||\overset{\circ}{\epsilon}||^2/(n-p_S)$. For $j \notin S$, we set $\overset{\circ}{\beta}_j=0$. (3iii)

Now $\overset{\circ}{\beta}_1$ is an estimate of β_1. And $\xi\overset{\circ}{\beta}$ predicts the η of (2) from its ξ.

The main performance measures of interest are $\mathrm{MSE}=E\{(\overset{\circ}{\beta}_1 - \beta_1)^2\}$ and $\mathrm{MSPE}=E\{(\eta - \xi\overset{\circ}{\beta})^2\}$, the mean square error of estimate and the mean square prediction error, respectively. These may be taken conditionally on X, or unconditionally (averaged over X).

We also consider a version of R^2. For any subset T of columns, let

$$\rho_T^2 = (\sum_{j\in T} \beta_j^2)/(\sigma^2 + \sum_{j=1}^{p} \beta_j^2), \tag{4}$$

the true R^2 for a model based on columns in T. Let

$$\phi^2 = E\{\rho_S^2\}. \tag{5}$$

The expectation is over S, the random set of selected columns in (3).

Empirical workers often neglect the randomness in S, treating $\overset{*}{\beta}$ and $\overset{*}{\sigma}^2$ as OLS estimators. In other words, they take the model to be

$$Y = X_S\beta + \epsilon$$

where the ϵ_i's are iid $N(0,\sigma^2)$ variables – but S is the result of the search procedure. Then they use the conventional OLS formulas for MSE, MSPE, and R^2. That is, they estimate the MSE of $\overset{*}{\beta}_1$ by

$$\text{naive MSE} \;=\; \overset{*}{\sigma}^2 \cdot \text{ the (1,1)-element of } (X_S^T X_S)^{-1}. \tag{6}$$

Likewise,

$$\text{naive MSPE} \;=\; \overset{*}{\sigma}^2 \cdot \{1 + \text{trace}\,(X_S^T X_S)^{-1}\}. \tag{7}$$

And ρ_S^2 is estimated by $\overline{R}^2$, where

$$1 - \overline{R}^2 \;=\; \frac{n}{n-p_S}(1 - R^2) \;=\; \overset{*}{\sigma}^2/(\|Y\|^2/n). \tag{8}$$

As will be seen, these estimators tend to be much too optimistic: in effect, they ignore the component of variance due to model selection.

Only the notation in (6-7-8) is unfamiliar. In the OLS context, $E\{\hat{\beta}_1 \,|\, X\}=\beta_1$. And $\text{var}\{\hat{\beta} \,|\, X\}=\sigma^2 \cdot (X^TX)^{-1}$ is estimated by putting $\hat{\sigma}^2$ in place of σ^2, giving (6). With respect to (7), if $\tilde{\beta}$ is any estimator for β based on X and Y,

$$E\{(\eta - \xi\tilde{\beta})^2|X\} \;=\; \sigma^2 + E\{\|\tilde{\beta} - \beta\|^2|X\}. \tag{9}$$

In the OLS case,

$$E\{\|\hat{\beta} - \beta\|^2|X\} \;=\; \sigma^2 \cdot \text{trace}\,(X^TX)^{-1}$$

and σ^2 is estimated by $\hat{\sigma}^2$. Formula (8) is close to standard, as in (Theil, 1971, p178): by (4), if $T = \{1, \cdots, p\}$,

$$1 - \rho_T^2 \;=\; \sigma^2/(\sigma^2 + \sum_{j=1}^{p}\beta_j^2)\,.$$

Numerator and denominator are estimated separately as $\hat{\sigma}^2$ and $\|Y\|^2/n$.

Coming now to the jackknife and cross validation, for each i let $Y^{(i)}$ and $X^{(i)}$ denote the result of deleting row i from the matrix. Let $\overset{*}{\beta}^{(i)}$ denote the estimator of β obtained by the screening process (3) applied to the ith reduced data set. Then

$$\text{jackknife MSE} \;=\; \frac{n-1}{n} \sum_{i=1}^{n} [\overset{*}{\beta}_1^{(i)} - \overset{*}{\beta}_1^{(-)}]^2 \tag{10}$$

where

$$\overset{*}{\beta}_1^{(-)} \;=\; \frac{1}{n} \sum_{i=1}^{n} \overset{*}{\beta}_1^{(i)}$$

(In principle, the jackknife is only considered to pick up the variance component of MSE.) For cross validation,

$$\text{cross validation MSPE} \;=\; \frac{1}{n} \sum_{i=1}^{n} (Y_i - \overset{*}{\hat{Y}}_i)^2 \tag{11}$$

where

$$\overset{*}{\hat{Y}}_i \;=\; (\text{row i of X}) \cdot \overset{*}{\beta}^{(i)}$$

and $\cdot$ stands for inner product. (Despite the notation, $\overset{*}{\hat{Y}} \neq X\overset{*}{\hat{\beta}}$.) In particular, the screening process is applied separately to each of the reduced data sets.

Psychologists often use the "cross-validated R^2:" in the present context, this may be taken as

$$(Y \cdot \overset{\curlyvee}{Y})^2 / \|Y\|^2 \, \|\overset{\curlyvee}{Y}\|^2 \tag{12}$$

and viewed as an estimate of ρ_S^2 in (4). Here, $\overset{\curlyvee}{Y}$ is defined as for (11).

Consider next the bootstrap. The idea is to estimate performance characteristics in a simulation model estimated from the data, and two choices present themselves for the parameters: using $\hat{\beta}$ to generate the starred data, or $\overset{\scriptscriptstyle\wedge}{\hat{\beta}}$. We elected to use $\hat{\beta}$, and found the bootstrap did not perform well: $\overset{\scriptscriptstyle\wedge}{\hat{\beta}}$ would make things even worse; indeed, $\overset{\scriptscriptstyle\wedge}{\hat{\beta}}$ can become in effect a self-fulfilling prophecy. Other choices present themselves for the explanatory variables: the bootstrap can be run conditionally by keeping X fixed and resampling the disturbances; or unconditionally, resampling X as well from its distribution, which is known in the present case. The conditional bootstrap seems more interesting, and turns out to perform better, so we report that. In principle, we view the conditional version of the bootstrap as estimating the conditional MSE or MSPE given X. Of course, eg, $E\{E[(\overset{\scriptscriptstyle\wedge}{\hat{\beta}}_1 - \beta_1)^2 \mid X]\} = E\{(\overset{\scriptscriptstyle\wedge}{\hat{\beta}}_1 - \beta_1)^2\}$. So, if all went well, the conditional bootstrap would also give nearly unbiased estimates of the unconditional MSE or MSPE. A third option – resampling rows – is not available in this problem: there is a high probability of getting fewer than 75 distinct rows in the resampling, so the rebuilt cross-product matrix will usually not be invertible. In any case, the empirical distribution of 100 data points in R^{75} is not a good estimate of the theoretical distribution.

To spell out the bootstrap procedure in more detail, given X and ϵ let $\hat{\beta}$ be the OLS estimate of β in (1). Let

$$Y^* = X\hat{\beta} + \epsilon^* \tag{13}$$

where ϵ^* is an $n \times 1$ vector of iid $N(0,\hat{\sigma}^2)$ variables. In principle, we also consider

$$\eta^* = \xi^*\hat{\beta} + \delta^* \tag{14}$$

where ξ^* is another $1 \times p$ row vector of iid $N(0,1)$ variables, and δ^* is $N(0,\hat{\sigma}^2)$. Pretend for a moment that $\hat{\beta}$ in the model (13) is an unknown parameter vector to be estimated by the selection procedure (3): run Y^* on X to get OLS estimates $\hat{\beta}^*$; let S^* be the set of significant columns, with $1 \in S^*$ by fiat; let $\overset{\scriptscriptstyle\wedge}{\hat{\beta}}^* = (X_{S^*}^T X_{S^*})^{-1} X_{S^*}^T Y^*$.

The bootstrap estimates of the performance measures are as follows:

$$\text{bootstrap MSE} = E_*\{(\overset{\scriptscriptstyle\wedge}{\hat{\beta}}_1^* - \hat{\beta}_1)^2\} \tag{15}$$

$$\text{bootstrap MSPE} = E_*\{(\eta^* - \xi^*\overset{\scriptscriptstyle\wedge}{\hat{\beta}}^*)^2\} = \hat{\sigma}^2 + E_*\{\|\overset{\scriptscriptstyle\wedge}{\hat{\beta}}^* - \hat{\beta}\|^2\} \tag{16}$$

$$\text{bootstrap } R^2 = E_*\{(\underset{j \in S^*}{\Sigma} \hat{\beta}_j^2)/(\sigma^2 + \Sigma_j \, \hat{\beta}_j^2)\}. \tag{17}$$

In these formulas, X and ϵ are held fixed. As will be seen, the bootstrap estimates of MSE and MSPE are too high. Paradoxically, so is the bootstrap R^2. References are given on the bootstrap, especially (Efron, 1979, 1982). For asymptotic theory, see (Beran, 1982), (Bickel and Freedman, 1981), (Freedman, 1981b); for applications, (Freedman and Peters, 1984abc).

2. Empirical results.

This section reports simulation results for the screening procedure $\overset{\scriptscriptstyle\wedge}{\hat{\beta}}$ defined by (3). The naive, bootstrap, and jackknife estimates of squared error will be compared, for estimation (MSE) and prediction (MSPE). The basic model is (1), with 100 rows and 75 columns, so

$n{=}100$ and $p{=}75$. And $\sigma^2{=}1$. Take $\beta_j{=}.2$ for $1\leq j\leq 25$ and $\beta_j{=}0$ for $26\leq j\leq 75$, so $\gamma{=}.2$ and $p_1{=}25$. For Table 1, we generated 100 basic data sets following (1): making the number of replicates equal to the number of rows was a matter of taste rather than necessity. The "true value" for $E\{(\hat{\hat{\beta}}_1 - \beta_1)^2\}$ is the empirical average

$$\frac{1}{100}\sum_{r=1}^{100}[\hat{\hat{\beta}}_1(r) - .2]^2$$

where $\hat{\hat{\beta}}_1(r)$ is the computed value of $\hat{\hat{\beta}}_1$ for the rth data set. As shown in the table, this average is .031. The SD of the 100 values $\{\hat{\hat{\beta}}_1(r) : r{=}1,...,100\}$ is quite large, .039. Still, the SE for the average is .0039. So the instability in the Monte Carlo is small. For the naive MSE, we report the average and SD of the 100 values $\hat{\sigma}^2(r) \cdot (1,1)$-element of $[X(r)_{S(r)}^T X(r)_{S(r)}]^{-1}$, with $1\leq r\leq 100$. As before, $\hat{\sigma}^2(r)$ is the value of $\hat{\sigma}^2$ for the rth data set, $X(r)$ is the rth matrix of explanatory variables, and $S(r)$ is the set of columns selected by procedure (3) applied to the rth data set. At .012, the naive MSE averages less than half what it should be. For the jackknife MSE, we report the average and SD of the 100 values

$$\text{jackknife MSE(r)} = \frac{n-1}{n}\sum_{i=1}^{n}[\hat{\hat{\beta}}_1^{(i)}(r) - \hat{\hat{\beta}}_1^{(-)}(r)]^2 \tag{18}$$

for $r{=}1,...,100$, which result from applying formula (10) to the rth data set. On average, the jackknife is too high by a factor of about 8. Whether viewed as estimating the conditional or unconditional MSE, the jackknife is not estimating it well.

Finally, for the bootstrap MSE, we report the average and SD of the 100 numbers generated by applying formula (15) to the rth data, for $r{=}1,...,100$. On average, the bootstrap is about 15% too high. And there is quite a lot of variability (from one data set to another) in the bootstrap estimate, as will be discussed later.

To approximate $E_*\{[\hat{\hat{\beta}}_1^*(r) - \hat{\beta}_1(r)]^2 \mid X(r)\}$ we generate 100 starred data sets according to (13), with $X(r)$ and $\hat{\beta}(r)$ in place of X and $\hat{\beta}$. (The equality of the number of replications in the various processes is still a matter of choice.) Specifically, for each r we generate 100 vectors of errors, each having 100 iid $N(0,\hat{\sigma}(r)^2)$ components. Corresponding to the sth vector $\epsilon(r,s)$ we make $Y(r,s){=}X(r)\hat{\beta}(r) + \epsilon(r,s)$, run $Y(r,s)$ on $X(r)$ according to the screening procedure (3), and come up with $\hat{\hat{\beta}}_1(r,s)$: the vector $Y(r,s)$ is $100{\times}1$ and the matrix $X(r)$ is $100{\times}75$. Then the bootstrap estimate for the MSE of $\hat{\hat{\beta}}_1$ given $X(r)$ is

$$E_*\{[\hat{\beta}_1^*(r) - \hat{\beta}_1(r)]^2 \mid X(r)\} \approx \frac{1}{100}\sum_{s=1}^{100}\{[\hat{\hat{\beta}}_1(r,s) - \hat{\beta}_1(r)]^2\} \tag{19}$$

The MSPE calculations are similar, and will not be recited in detail. On average, cross-validation does quite well, but the bootstrap is nearly 30% too high. Both show a lot of variability. In the R^2-column, the "true value" is an approximation to $\phi^2{=}E\{\rho_S^2\}$, obtained by averaging the values for the 100 data sets. As can be seen, the naive estimate is on average more than double the true value, and the bootstrap is worse. Cross validation is low, also by a factor of nearly 2.

Some benchmarks are shown at the bottom of the table. An old-fashioned statistician might elect to estimate β_1 by regressing Y on the first column of X, ie, neglecting the covariates: this procedure, $\hat{\beta}_{\text{no adjust}}$, does quite well, with a variance of .021. Another statistician might put in all the covariates: var $(\hat{\beta}_{\text{OLS}}){=}.042$. This is not so good.

The calculation for the variance of $\hat{\beta}_1$:

var $(\hat{\beta}_1 \mid X) = \sigma^2 V_1$,

Table 1. Simulation results for the jackknife and bootstrap applied to the screening estimator $\overset{*}{\beta}$. The model specification: $\sigma^2=1$, n=100, p=75, $p_1=25$, $\gamma=.2$

	Estimates of MSE		Estimates of MSPE		Estimates of R^2	
	ave	SD	ave	SD	ave	SD
true	.031	.039	2.79	.63	.243	.064
naive	.012	.003	1.18	.25	.561	.128
jackknife	.233	.113	*	*	*	*
cross validation	*	*	2.70	.57	.152	.107
bootstrap	.036	.015	3.56	.84	.677	.088

$$\text{var}\,(\hat{\beta}_{\text{no adjust}}) = .021 \qquad \text{MSPE}\,(\overline{Y}) = 2.01$$

$$\text{var}\,(\hat{\beta}_1) = .042 \qquad \text{MSPE}\,(\xi\hat{\beta}) = 4.12$$

where V_1 is the (1,1)-element of $(X^TX)^{-1}$ and is distributed as $1/\chi^2_{n-p+1}$ with expectation $1/n-p-1$. The computation for $\text{var}\,(\hat{\beta}_{\text{no adjust}})$ is similar, except that σ^2 must be revised upward to $1 + 24 \times (.2)^2 = 1.96$ to account for the omitted covariates, and p=1 not 75.

Why is OLS so bad? In principle, covariate adjustment should improve precision. But there is a tradeoff, since adding variables degrades the quality of the coefficient estimates: roughly speaking, adding an unnecessary variable is like throwing away a data point. See (Breiman and Freedman, 1983) or (Eaton and Freedman, 1982). By comparison, the screening procedure shrinks the estimated coefficients towards 0, and this improves the accuracy relative to OLS. In Table 1, however, the best strategy is still not to adjust at all.

Similar benchmarks are shown for the prediction problem: predicting $\eta=\xi\beta + \delta$ by $\overline{Y}$, ie, ignoring the covariates, has an MSPE of 2.01. In the circumstances, this is the best of the procedures we consider. Predicting η by $\xi\hat{\beta}$, the conventional OLS strategy using all the covariates, has an MSPE of 4.12. This is the worst.

How sensitive are these results to the selected value for β? To address this question we set the common value of β_i for $1\leq i\leq 25$ to $\gamma=.1$, .5 and 1.0 as well as to .2. The results are not qualitatively different, except that for large values of γ the R^2's are all close to 1, and the balance tilts toward covariate adjustment. Of course, the independence assumption matters too.

The difficulties in Table 1 are mainly due to the fact that p/n is near 1. To illustrate the point, consider Table 2, where n=100 but p is reduced to 10; the first five β's are set at .2, the others at 0. When p is much smaller than n, the impact of the screening process (3) is small, since at most p/(n-p) of the degrees of freedom for error are being juggled. The naive, bootstrap and cross-validation procedures all give similar results for MSE and MSPE, although R^2 is still hard to estimate.

The jackknife estimate is still too big, by about 50%. We have no explanation to offer; on the other hand, we never understood why the jackknife was supposed to work, except as an approximation to the bootstrap (Efron, 1982, Chapter 6; and see Chapter 4 on bias in the jackknife). We also tried the jackknife on OLS, ie, to estimate $\text{var}\,(\hat{\beta}_1)$. Somewhat to our surprise, the jackknife was still about 20% too high; on the other hand, as theory predicts, the bootstrap came in right on the money.

Table 2. Simulation results for the jackknife and bootstrap applied to the screening estimator $\hat{\hat{\beta}}$. The model specification: $\sigma^2{=}1$, $n{=}100$, $p{=}10$, $p_1{=}5$, $\gamma{=}.2$

	Estimates of MSE		Estimates of MSPE		Estimates of R^2	
	ave	SD	ave	SD	ave	SD
true	.0097	.0125	1.11	.062	.139	.024
naive	.0108	.0021	1.05	.153	.185	.066
jackknife	.0156	.0068	*	*	*	*
cross validation	*	*	1.14	.184	.103	.071
bootstrap	.0116	.0029	1.15	.170	.211	.068

$$\text{var}\,(\hat{\beta}_{\text{no adjust}}) \;=\; .012 \qquad \text{MSPE}\,(\overline{Y}) \;=\; 1.21$$
$$\text{var}\,(\hat{\beta}_1) \;=\; .011 \qquad \text{MSPE}\,(\xi\hat{\beta}) \;=\; 1.11$$

3. Reasons for bootstrap failure.

Although $\hat{\beta}$ is an unbiased estimator of β, there is bias in $||\hat{\beta}||^2$, conditionally or unconditionally:

$$E\{||\hat{\beta}||^2|X\} \;=\; ||\beta||^2 + \sigma^2 \cdot \text{trace}\,(X^T X)^{-1}$$
$$E\{||\hat{\beta}||^2\} \;=\; ||\beta||^2 + p\sigma^2/(n{-}p{-}1).$$

In other words, the bootstrap model (13) starts from a parameter vector with a much inflated length. In Table 1, for example, $||\beta||^2{=}1$ and $p\sigma^2/(n{-}p{-}1) \approx 3$. This explanation for bootstrap failure suggested deflating $\hat{\beta}$ by the appropriate factor (namely, $[1 + \hat{\sigma}^2 \cdot \text{trace}\,(X^T X)^{-1}/||\hat{\beta}||^2]^{1/2}$) to get its length about right, before resampling. For the model in Table 1, length adjustment does bring the bootstrap into better line: see Table 3.

We also tried a model with $\beta_j{=}.2$ for $j{=}1,...,75$ so $p_1{=}75$. See Table 4. In this case, adjustment makes things worse on MSE and MSPE: indeed, the raw bootstrap is already biased downward. For R^2, the adjustment helps. We do not recommend length adjustment without further analysis.

Table 3. Simulation results for the raw and length-adjusted bootstrap on the screening estimator $\hat{\hat{\beta}}$: the model specification is as in Table 1.

	Estimates of MSE		Estimates of MSPE		Estimates of R^2	
	ave	SD	ave	SD	ave	SD
true	.031	.039	2.79	.63	.243	.064
raw bootstrap	.036	.015	3.56	.84	.677	.088
adjusted bootstrap	.025	.011	2.48	.50	.167	.157

Table 4. Simulation results for the raw and length-adjusted bootstrap on the screening estimator $\overset{\circ}{\hat{\beta}}$. Model specification: $\sigma^2=1$, n=100, p=75, $p_1=75$, $\gamma=.2$

	Estimates of MSE		Estimates of MSPE		Estimates of R^2	
	ave	SD	ave	SD	ave	SD
true	.057	.087	4.63	.81	.353	.065
raw bootstrap	.045	.022	4.21	.85	.800	.057
adjusted bootstrap	.037	.019	3.48	.56	.352	.154

With respect to the model in Table 1, denote the MSE given X by

$$\Phi(\beta,\sigma,X) \; = \; E\{(\overset{\circ}{\hat{\beta}}_1 - \beta_1)^2 \mid X\}. \tag{20}$$

The bootstrap approximates $\Phi(\beta,\sigma,X)$ by $\Phi(\hat{\beta},\hat{\sigma},X)$, and in effect our tables on MSE compare $E\{\Phi(\beta,\sigma,X)\}$ to $E\{\Phi(\hat{\beta},\hat{\sigma},X)\}$. In principle, Φ depends on all the coordinates of β, and in this respect screening differs from OLS, where $E\{(\hat{\beta}_1 - \beta_1)^2\}$ does not depend on β.

One explanation for bootstrap failure is strong nonlinear dependence of Φ on β. About the strongest we found was on $||\beta||^2$. To represent the data more conveniently, let

$$\Phi(\beta,\sigma) \; = \; E\{\Phi(\beta,\sigma,X)\} \; = \; E\{(\overset{\circ}{\hat{\beta}}_1 - \beta_1)^2\} \; . \tag{21}$$

This is the unconditional MSE. Figure 1 shows a plot of $\Phi(\beta,\sigma)$ against $||\beta||^2$ or σ^2. (The values of β and σ were drawn as a sample from the OLS distribution of $\hat{\beta}$ and $\hat{\sigma}$; computationally, we estimated $\Phi(\hat{\beta},\hat{\sigma})$ by running the unconditional bootstrap.) By regression,

$$\Phi(\hat{\beta},\hat{\sigma}) \; = \; .0035 \times ||\hat{\beta}||^2 + .026 \times \hat{\sigma}^2 + \text{residual}, \qquad R^2 = .70 \tag{22}$$

Since $||\hat{\beta}||^2$ tends to be too big, this does inflate the bootstrap estimate of MSE, as indicated at the beginning of the section – for the model in Table 1.

Switching now from estimation to prediction, a heuristic explanation for the bias in the bootstrap R^2 and MSPE runs as follows. Keeping σ^2 fixed, R^2 measures how big the β's are, and the MSPE measures how well they are estimated. The $\hat{\beta}$'s tend to be too big, inflating R^2. On the other hand, when a big $\hat{\beta}_i$ is estimated as 0 by the corresponding bootstrap $\overset{\circ}{\hat{\beta}}_i{}^*$, that is a big error.

To quantify the effect, for any subset H of columns let $H^1=H$ while H^0 is the complement of H. Let J be the set of columns j with $1 \leq j \leq 25$, so $\beta_j=\gamma$ is positive for $j \in J^1 = J$ while $\beta_j=0$ for $j \in J^0$. Recall the set S of selected columns from (3) and S^* from the discussion before (15). For a,b=0 or 1 let

$$E_{ab} \; = \; E\left\{\Sigma_{j \in J^a \cap S^b} (\overset{\circ}{\hat{\beta}}_j - \beta_j)^2\right\} \quad \text{and} \quad E_{ab}^* \; = \; E_*\left\{\Sigma_{j \in J^a \cap S^{*b}} (\overset{\circ}{\hat{\beta}}_j{}^* - \hat{\beta}_j)^2 \mid X\right\} \; . \tag{23}$$

Starting from equation (9),

$$\text{MSPE} \; = \; \sigma^2 + E_{11} + E_{01} + E_{10} + E_{00}.$$

Likewise from (16),

$$\text{bootstrap MSPE} \; = \; \hat{\sigma}^2 + E_{11}^* + E_{01}^* + E_{10}^* + E_{00}^*.$$

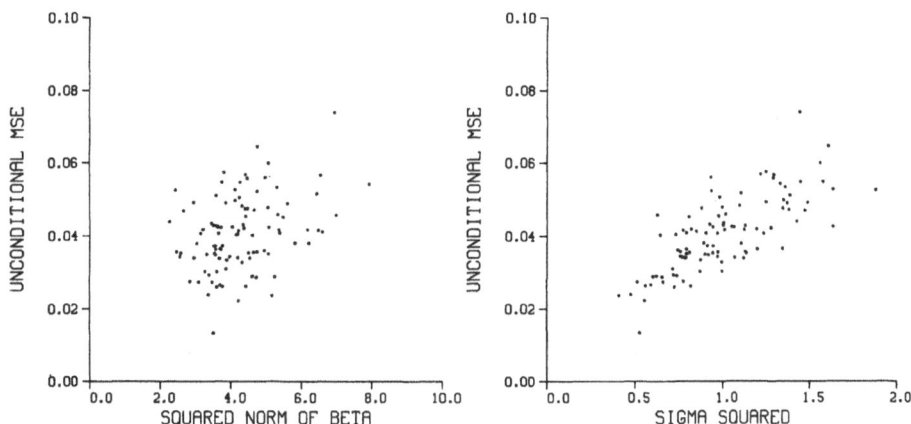

Figure 1. Plot of unconditional MSE against $||\beta||^2$ and σ^2, for a sample of β's and σ's.

The average values for these 5 components of MSPE are shown in Table 5. As will be seen, most of the bias in the bootstrap can be accounted for by the last row in the table corresponding to j's which have $\beta_j = 0$ in the true model and are screened out of the bootstrap model: $26 \leq j \leq 75$ and $j \notin S^*$. Of course, such j's have $\hat{\beta}_j \neq 0$, and that is the problem in Table 1. Indeed, E_{00} is necessarily 0 while E_{00}^* is quite positive. By contrast, $E_{00} = E_{00}^* = 0$ in Table 4, where the bootstrap is biased downward.

Table 5. Simulation results for the components of MSPE and bootstrap MSPE: the model specification is as in Table 1.

	True	Bootstrap
variance	1.00	0.96
$1 \leq j \leq 25$ and j selected	0.40	0.53
$26 \leq j \leq 75$ and j selected	0.88	0.92
$1 \leq j \leq 25$ and j not selected	0.51	0.43
$26 \leq j \leq 75$ and j not selected	0.00	0.72
total	2.79	3.56

4. Other findings.

a) *The conditional MSE.* Table 1 shows the unconditional average and SD of $(\hat{\hat{\beta}}_1 - \beta_1)^2$, as $.031 \pm .039$. For each of the 100 data sets $r=1,...,100$ in the simulation, consider the conditional mean square error $E\{(\hat{\hat{\beta}}_1 - \beta_1)^2|X(r)\}$. To estimate this conditional expectation, we generated for each r a set of 100 vectors of errors, each vector having 100 iid N(0,1) components. Corresponding to the sth vector $\epsilon(r,s)$, we made $Y(r,s)=X(r)\beta + \epsilon(r,s)$ and applied the screening process (3) to $Y(r,s)$ and $X(r)$, winding up with $\hat{\hat{\beta}}(r,s)$. The conditional MSE of $\hat{\hat{\beta}}_1$ given $X(r)$ can now be estimated as

$$ \text{MSE}(r) \;=\; \frac{1}{100} \sum_{s=1}^{100} [\hat{\hat{\beta}}_1(r,s) - .2]^2. \tag{24} $$

These 100 conditional MSE's averaged out to .028, with an SD of .0084. The difference between .028 and .031 $\approx \text{E}\{(\hat{\hat{\beta}}_1 - \beta_1)^2\}$ is sampling error, and the .028 is more reliable. Indeed, the difference between .0084 and .039 $\approx$ SD of $(\hat{\hat{\beta}}_1 - \beta_1)^2$ shows how conditioning on X dramatically reduces the variability in $(\hat{\hat{\beta}}_1 - \beta_1)^2$.

For each data set r, we previously computed in (19) the bootstrap estimate for the MSE of $\hat{\hat{\beta}}_1$ given $X(r)$, starting from $\hat{\beta}$ and $\hat{\sigma}$ rather than β and σ. A scatter plot of the bootstrap estimate against the conditional MSE across data sets is shown in the left hand panel of Figure 2; a similar plot for the jackknife is shown at the right. As will be clear, the bootstrap is consistently too high, by a little. The jackknife is an order of magnitude too big. Furthermore, the R^2 for the bootstrap is only 0.36; for the jackknife, 0.19. In addition to other troubles, these methods cannot discriminate very well between informative and uninformative data sets. (There is no real attenuation due to imprecision in the Monte Carlo.)

b) *Outliers.* As will be clear from Figure 2, the jackknife estimate has quite a long right hand tail. On the log scale in the right hand panel of Figure 3, the bias is still plain to see. The left hand panel gives a scatter plot for the log bootstrap; this looks quite normal, but R^2 is only 0.25. By regression,

$$ \text{bootstrap estimate} \;=\; .43 \times (\text{true MSE given X})^{.71} \times \text{residual factor} \tag{25} $$

The small bias in the bootstrap can still be discerned; a majority of the points are above the 45-degree line.

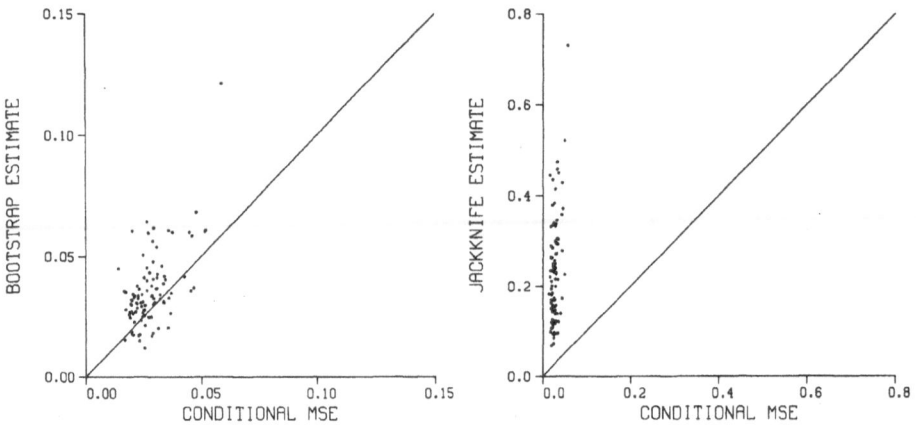

Figure 2. The left hand panel plots for each of 100 data sets the bootstrap estimate of mean square error against its true value (conditional on X). The right hand panel does the same for the jackknife. The scales differ. The 45-degree line is plotted for reference.

Table 6. Root mean square error for various estimates of MSE and MSPE: the model specification is as in Table 1.

	Estimates of MSE	Estimates of MSPE
naive	.018	1.77
jackknife	.232	*
cross validation	*	0.81
bootstrap	.014	1.36

c) *RMS error*. As another measure for the accuracy of the naive, jackknife and bootstrap MSE, we took the root mean square difference between each of these estimates and the true MSE conditional on X, over the 100 data sets in the simulation discussed in paragraph a). The results are shown in Table 6. The bootstrap is only a little better than the naive estimate: increased variability trades off against decreased bias. Table 6 also shows the results for MSPE. Here, the cross validation estimator is superior. The bootstrap estimates are not bad, on average (Table 1). But they are quite noisy: that is the message of this paragraph.

d) *Bias in the screening estimator*. When averaged over X, the screening estimator $\overset{*}{\hat{\beta}}_1$ is unbiased by symmetry. Indeed, the first column of X is entered automatically; now project into its orthocomplement and use rotational invariance. However, $\overset{*}{\hat{\beta}}_1$ is conditionally biased given X. For the simulation discussed in paragraph a), $E(\overset{*}{\hat{\beta}}_1|X)$ averaged .20 with an SD of .038; the SD measures the conditional bias for a typical X as about 20% of the true value.

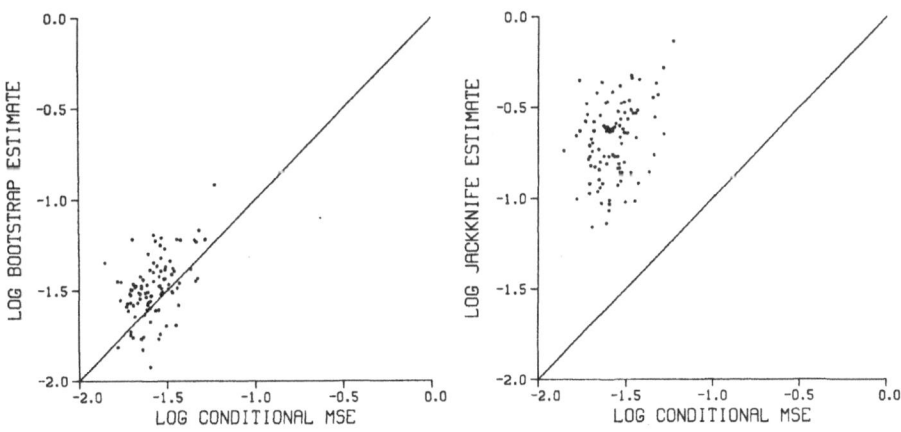

Figure 3. The left hand panel plots for each of 100 data sets the log of the bootstrap estimate of MSE against the log of the true value conditional on X. The right hand panel does the same for the jackknife. The logarithms are to base 10.

For columns 2 through 25, the screening estimator is biased toward 0, conditionally or unconditionally. For example, a hundred values of $\overset{*}{\hat\beta}_2$ averaged .14 with an SD of .025 and an SE of .0025; again, the true value is .20.

e) *The effect of refitting.* Any two columns of X are nearly orthogonal, so the effect of refitting in (3) should be minimal: ie, $\overset{*}{\hat\beta}_j \approx \hat\beta_j$ for j ∈ S. To test this idea, we compared $\Sigma_{j\in S}\,\overset{*}{\hat\beta}_j{}^2$ with $\Sigma_{j\in S}\,\hat\beta_j{}^2$. In the simulation for Table 1, the first sum averages 2.2 and the second, 3.3. Refitting matters; there are a lot of pairs of columns and the non-orthogonality mounts up.

f) *F-tests for omitted variables.* It has been suggested that our results are due to model mis-specification, which could be detected by a routine F-test. We disagree. The explanation for our results is chance capitalization: data-snooping distorts conventional measures of goodness-of-fit. Indeed, the F-test cannot detect the miss-specification. To illustrate the point, consider the simulation for Table 1. This involved generating 100 data sets following the model (1); and for each, performing the screening operation (3), leading to a set S of selected columns. For each data set, we ran a naive F-test for adding *en bloc* the columns outside S. On the average, the F-statistic was .9, with 49.4 degrees of freedom in the numerator and 25 in the denominator. (Also see Table 7 below.) This would only confirm the value of the screening procedure. Of course, it is misleading to make F-tests this way, treating S as given rather than the result of data-snooping.

g) *How many variables get into the second pass?* In the simulation for Table 1, the coefficients of the first 25 columns were set to a common positive value; these will be called 1-columns. The remaining 50 columns had coefficients set to 0, and will be referred to as 0-columns. Let N_1 be the number of 1-columns which got into the second-pass regression. Likewise, let N_0 be the number of 0-columns which entered the second-pass regression. The bootstrap analogs will be denoted by stars: thus, N_0^* is the number of columns with $25 \le j \le 75$ which entered the second-pass bootstrap regression. (Of course, $\hat\beta_j \ne 0$ even for the 0-columns.)

Means for these N's are shown in Table 7, for a simulation involving 100 data sets. For example, we expect $.25 \times 50 = 12.5$ of the 0-columns to get in, and on the average 13.4 did: the difference is sampling error. (Since X is not exactly orthogonal, the $\hat\beta$'s are dependent, and the variability in N_0 is appreciably greater than binomial.)

On the average, 12.2 of the 1-columns got into the second-pass regression. This is only 49% of the 1-columns, which may seem disappointing, but in the present context even a test of size 25% does not have much power. The bootstrap estimates this quite well: $E(N_1^*) = 13.8$. However, the bootstrap badly over-estimates the number of 0-columns: 21.4 versus 13.4. This is because the $\hat\beta$'s tend to be too large, so the $\hat\beta^*$'s are more likely to be significant.

Table 7. Simulation results for the number of variables entering the second pass: the model specification is as in Table 1.

	1-columns	0-columns	total
true	12.2	13.4	25.6
bootstrap	13.8	21.4	35.2

Dijkstra (as reported in these proceedings) had a sharper result for a smaller model. To replicate his work, we repeated our simulation for a model with five 1-columns and five 0-columns. The results are shown in Table 8: the bootstrap is over 50% too high on the 0-columns.

A small theoretical calculation might clarify matters. Consider the very simple regression model

$$Y_i = \beta x_i + \epsilon_i \tag{26}$$

where the ϵ_i are iid N(0,1) for $i=1,...,n$. Here, β is just a number. The x's are deterministic, and normalized so $\Sigma_1^n x_i^2 = n$. Fix a critical value c and let

$$\Phi(\beta) = \Pr\{|\hat{\beta}| > c/\sqrt{n}\}. \tag{27}$$

Of course, this Φ can be computed exactly from the normal distribution, since $\sigma^2=1$ is given:

$$\Phi(\beta) = \Pr\{|\beta\sqrt{n} + Z| > c\} \tag{28}$$

where Z is N(0,1). Indeed, $\hat{\beta}$ is distributed as $\beta + (Z/\sqrt{n})$.

Now we try to estimate Φ by the bootstrap:

$$\Phi(\hat{\beta}) = \Pr\{|\hat{\beta}\sqrt{n} + Z'| > c\} \tag{29}$$

where Z' is an independent N(0,1) variable, and Z is held fast. Finally,

$$E\{\Phi(\hat{\beta})\} = \Pr\{|\beta\sqrt{n} + Z + Z'| > c\} \tag{30}$$

where Z and Z' both vary. If β is of order $1/\sqrt{n}$ or smaller, the bootstrap will fail: $Z + Z'$ has fatter tails than Z, by a lot. If $\beta\sqrt{n} \to \infty$, then $\Phi(\beta)$ and $E\{\Phi(\hat{\beta})\}$ will both approach 1, but at different rates.

Table 8. Simulation results for the number of variables entering the second pass. The model specification: $\sigma^2=1$, n=100, p=10, $p_1=5$, $\gamma=.2$

	1-columns	0-columns	total
true	4.2	1.3	5.5
bootstrap	3.9	2.1	6.0

5. Computational details.

The program was written in FORTRAN, using LINPAK for the matrix algebra. The computations were done on a CRAY. Those for the model in Table 1, for example, took 10 minutes of CPU time. Among other things, there were a hundred 75×75 matrices to invert, and upwards of 50,000 regressions to run. (Cross-validation was done by updating $X^T X$: see Efron, 1982, p18). Some of calculations were replicated on a SUN workstation, in FORTRAN and in S. A few of them were replicated in True BASIC on a PC-XT. We therefore have some degree of confidence in the code. Too, exact distributions for many of the intermediate results can be computed and checked against observations. On the whole, this worked out quite well; there were a few small but highly significant anomalies. Of course, we are pushing the random number generator quite hard: Table 1 involves over a million calls.

14

6. Summary and conclusions.

In our simulations, when the number of variables is relatively large the bootstrap and particularly the jackknife have some trouble in dealing with uncertainty created by variable selection. It may not be possible on the basis of such techniques to develop a model and calculate its performance characteristics on the same data set. This would have gloomy implications for many kinds of modeling. Of course, an investigator can always develop the model on one data set and test it on another: replication is always a good idea.

In the classical setup, given some type of relationship among variables expressed in a well-specified statistical model, it is possible to estimate parameters or make predictions from a data set and put margins of error on the results. If you know what to look for, there is a way to find it. On the other hand, given any statistical procedure there will always be some kinds of relationships which will not be detected by that procedure. And someone who uses a variety of statistical procedures, taking many cuts at the data, is almost bound to find structure even when none exists. That is the trouble with data-snooping.

To illustrate the point that given some style of analysis there will be structure which escapes it, take linear regression analysis. Consider the time series x_t plotted against time $t=1, \cdots ,50$ at the left in Figure 4. This looks like pure noise, and fitting $x_t = a + bt + e_t$ isolates no trend. On the other hand, plotting x_t against x_{t-1} at the right shows this series to be perfectly deterministic: $x_t = f(x_{t-1})$, where

$$f(x) = 2x \qquad \text{for } 0 \leq x \leq 1/2$$
$$= 2 - 2x \quad \text{for } 1/2 \leq x \leq 1.$$

A major part of the problem in applications is the curse of dimensionality: there is a lot of room in high-dimensional space. That is why investigators need model specifications tightly deri-

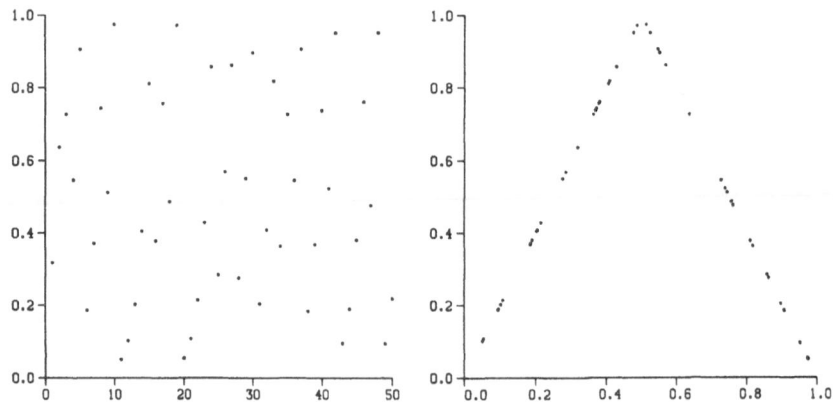

Figure 4. At the left, a time series with no linear regression structure. At the right, plotting x_t against x_{t-1} reveals the structure.

ved from good theory. We cannot expect statistical modeling to perform at all well in an environment consisting of large, complicated data sets and weak theory. Unfortunately, at present that describes many applications. References are given on modeling issues, eg, (Achen, 1982), (Baumrind, 1983), (Daggett and Freedman, 1985), (de Leeuw, 1985), (Freedman, 1985, 1986), (Freedman-Rothenberg-Sutch, 1983), (Hendry, 1980), (Leamer, 1983), (Ling, 1983), (McNees, 1986), (Zarnowitz, 1979).

Disclosures

Rudy Beran remarks that chance capitalization is a problem, even for bootstrap studies of chance capitalization. In principle, this is certainly right. However, in this paper we took our own advice about replication. We spent several months on free-style data snooping. Then we wrote a draft of the paper, with blank spaces for all the empirical numbers. Then we made a fresh set of computer runs and filled in those blanks. Finally, we ate all the words that had to be eaten.

Acknowledgements

We would like to thank Theo Dijkstra for his work in putting these proceedings together. He, Rudy Beran, Lincoln Moses and Jamie Robins made useful comments. Ani Adhikari provided lots of last-minute technical help. Our research was partially supported by NSF Grant DMS86-01634.

References

Achen, C. (1982). *Interpreting and using regression.* Beverly Hills, Calif.: Sage.

Baumrind, D. (1983). Specious causal attribution in the social sciences: the reformulated stepping-stone theory of heroin use as exemplar. *J. Pers. Soc. Psych.*, **45**, 1289-98.

Beran, R. (1984). Jackknife approximations to bootstrap estimates. *Ann. Statist.*, **12**, 101-118.

Bickel, P. and Freedman, D. (1981). Some asymptotic theory for the bootstrap. *Ann. Statist.*, **9**, 1196-1217.

Breiman, L. and Freedman, D. (1983). How many variables should be entered in a regression equation? *J. Am. Stat. Assoc.*, **78**, 131-136.

Daggett, R. and Freedman, D. (1985). Econometrics and the law: a case study in the proof of antitrust damages. In L. LeCam and R. Olshen (Eds.), *Proceedings of the Berkeley Conference in Honor of Jerzy Neyman and Jack Kiefer*, Vol I, 126-75. Belmont, Calif.: Wadsworth.

de Leeuw, J. (1985). Review of books by Long, Everitt, Saris and Stronkhorst. *Psychometrika*, **50**, 371-5.

Eaton, M. and Freedman, D. (1982). A remark on adjusting for covariates in multiple regression. Technical Report No. 11, Department of Statistics, University of California, Berkeley.

Efron, B. (1979). Bootstrap methods: another look at the jackknife. *Ann. Statist.*, **7**, 1-26.

Efron, B. (1982). *The Jackknife, the Bootstrap, and Other Resampling Plans.* Philadelphia: SIAM.

Freedman, D. (1981a). Some pitfalls in large econometric models: a case study. *J. Bus.* **54**, 479-500.

Freedman, D. (1981b). Bootstrapping regression models. *Ann. Statist.*, **9**, 1218-1228.

Freedman, D. (1983). A note on screening regression equations. *Am. Stat.*, **37**, 152-5.

Freedman, D. (1985). Statistics and the scientific method. In W. Mason and S. Fienberg (Eds.), *Cohort Analysis in Social Research: Beyond the Identification Problem*, 345-390 (with discussion). New York: Springer.

Freedman, D. (1986). As others see us: a case study in path analysis. Technical report, Department of Statistics, University of California, Berkeley. To appear in *J. Ed. Stat.*.

Freedman, D. and Navidi, W. (1986). Regression models for adjusting the 1980 Census. *Stat. Sci.*, **1**, 1-39.

Freedman, D. and Peters, S. (1984a). Some notes on the bootstrap in regression problems. *J. Bus. Econ. Stat.*, **2**, 406-409.

Freedman, D. and Peters, S. (1984b). Bootstrapping a regression equation: some empirical results. *J. Am. Stat. Assoc.*, **79**, 97-106.

Freedman, D. and Peters, S. (1984c). Bootstrapping an econometric model: some empirical results. *J. Bus. Econ. Stat.*, **2**, 150-8.

Freedman, D. and Peters, S. (1985). Using the bootstrap to evaluate a forecasting equation. *J. Forecasting*, **4**, 251-262.

Freedman, D., Rothenberg, T. & Sutch, R. (1983). On energy policy models. *J. Bus. Econ. Stat.*, **1**, 24-36. (With discussion.)

Gong, G. (1986). Cross-validation, the jackknife, and the bootstrap: excess error estimation in forward logistic regression. *J. Am. Stat. Assoc.*, **81**, 108-113.

Hendry, D. (1980). Econometrics – alchemy or science? *Econometrica*, **7**, 387-406.

Leamer, E. (1983). Taking the con out of econometrics. *Am. Econ. Rev.*, **73**, 31-43.

Ling, R. (1983). Review of *Correlation and Causation* by Kenny. *J. Am. Stat. Assoc.*, **77**, 489-91.

Lovell, M. (1983). Data mining. *Rev. Econ. Statist.*, **LXV**, 1-11.

McNees, S.K. (1986). Forecasting accuracy of alternative techniques: a comparison of US macroeconomic forecasts. *J. Bus. Econ. Stat.*, **4**, 5-24. (With discussion.)

Theil, H. (1971). *Principles of Econometrics.* New York: Wiley.

Zarnowitz, V. (1979). An analysis of annual and multiperiod quarterly forecasts of aggregate income, output, and the price level. *J. Bus.*, **52**, 1-34.

DATA-DRIVEN SELECTION OF REGRESSORS AND THE BOOTSTRAP

T.K. Dijkstra and J.H. Veldkamp
University of Groningen
Institute of Econometrics
Groningen, The Netherlands

1. INTRODUCTION

Consider the classical linear model with n observations, a fixed design matrix X, and i.i.d. Gaussian residuals with zero mean and positive variance. Suppose it is believed that some of the columns of X may be redundant, but it is not known which. Given the data a model search is carried out using the following criterion:

$$RSS(i_1, i_2, \ldots, i_p) + \alpha p s^2 \tag{1}$$

Here RSS(.) designates the residual-sum-of-squares obtained by regression on the p X-columns indexed by $i_1, i_2, \ldots, i_p$; s^2 is the usual unbiased estimate of the residual variance obtained by fitting all columns; and α is a positive number, often 2 but $\alpha = 1$ or $\alpha = \log n$ have also surfaced in the literature. The value of (1) is calculated for those sets of columns one is willing to consider and that model is chosen for which the criterion is minimal.

This procedure in which a compromise between "lack-of-fit" and "model-complexity" is minimized has received considerable attention, see e.g. Bhansali and Downham (1977), Atkinson (1980,1981), Geweke and Meese (1981), Leamer (1983), and the references therein. For fixed α it is well-known that a false model has, asymptotically, zero probability of minimizing (1) if at least one of the alternatives is a true model. In other words, in sufficiently large samples one will be choosing from true models only. The asymptotic probability of picking the simplest true model can differ substantially from 1 however. If α is allowed to increase with n at not too fast a rate, $\alpha = \log n$ will do, then we will be almost certain in sufficiently large samples to find the simplest true model.

A problem which has not received due attention is how to assess the reliability of coefficient estimates when a model search has taken place.

We will discuss this in some detail in section 5. First, however, we
will collect some information about the sampling distribution of coefficient estimators for a simple but interesting case. Section 2 is
devoted to model selection probabilities. Section 3 describes exact and
asymptotic distributions, conditional on the choice made as well as
unconditional. In section 4 we test whether a version of the bootstrap
can produce meaningful information about selection probabilities and
unconditional distributions. The reader is assumed to be familiar with
the bootstrap (or the "empirical monte carlo method" as we prefer to
call it); basic references are Efron (1979,1982) and Freedman (1981).

2. SELECTION PROBABILITIES

Suppose X has only two linearly independent columns. $X = [x_1, x_2]$.
Let $y = X\beta + \epsilon$ with $\epsilon \sim N_n(0, o^2 I)$. And suppose it is desired to choose
one hypothesis through minimization of (1) from the following four
alternatives: y is white noise, only x_1 is relevant, only x_2 has a nonzero coefficient, both x's contribute to the explanation of y. In this
section we will give some information about the probabilities, denoted
by p_0, p_1, p_2 and p_{12} resp., of ending up with one of the alternatives
mentioned. We will limit ourselves to the case where $\beta_2 = 0$, so
$y = \beta_1 x_1 + \epsilon$.

Let r be defined by $r \|x_1\| \cdot \|x_2\| = x_1' x_2$ and let

$$b \equiv (X'X)^{-1} X' y$$

$$b_{1.} \equiv (x_1' x_1)^{-1} x_1' y \qquad\qquad (2)$$

$$b_{.2} \equiv (x_2' x_2)^{-1} x_2' y$$

So

$$\frac{\|x_1\|}{o}(b_{1.} - \beta_1) = \frac{x_1' \epsilon}{o \|x_1\|} \equiv z_1$$

$$\frac{\|x_2\|}{o}(b_{.2} - r\beta_1 \frac{\|x_1\|}{\|x_2\|}) = \frac{x_2' \epsilon}{o \|x_2\|} \equiv z_2 \qquad\qquad (3)$$

$$\frac{\|x_{1.2}\|}{o}(b_1 - \beta_1) = \frac{z_1 - r z_2}{\sqrt{(1-r^2)}} \quad ; \quad \frac{\|x_{2.1}\| b_2}{o} = \frac{z_2 - r z_1}{\sqrt{(1-r^2)}}$$

where $\begin{bmatrix} z_1 \\ z_2 \end{bmatrix} \sim N_2(0, \begin{bmatrix} 1 & r \\ & 1 \end{bmatrix})$ and $x_{i.j} \equiv [I - x_j(x_j'x_j)^{-1}x_j']x_i$.

We also have the following decompositions

$$y = b_{1.}x_1 + e_1$$

$$y = b_{.2}x_2 + e_2 \qquad\qquad (4)$$

$$y = b_{1.}x_1 + b_2 x_{2.1} + e$$

$$y = b_1 x_{1.2} + b_{.2}b_2 + e$$

where $e \equiv y - Xb$, e_1 and e_2 are defined implicitly. Now consider p_0, which equals

$$Pr\{y'y < e_1'e_1 + \alpha s^2, \ e_2'e_2 + \alpha s^2, \ e'e + 2\alpha s^2\} \qquad\qquad (5)$$

Clearly, $p_0 \leqq Pr\{y'y < e_1'e_1 + \alpha s^2\} = Pr\{\dfrac{b_{1.}^2 \|x_1\|^2}{\sigma^2} < \alpha \dfrac{s^2}{\sigma^2}\} =$

$= Pr\{\chi^2(1, \dfrac{\beta_1^2 \|x_1\|^2}{\sigma^2}) < \alpha \dfrac{\chi^2(n-2,0)}{n-2}\}$ which tends to zero under the usual

assumptions concerning the design matrix X. Similarly, p_2 is bounded by

$$Pr\{e_2'e_2 + \alpha s^2 < e'e + 2\alpha s^2\} = Pr\{\dfrac{b_1^2 \|x_{1.2}\|^2}{\sigma^2} < \alpha \dfrac{s^2}{\sigma^2}\} =$$

$Pr(\chi^2(1, (1-r^2)\dfrac{\beta_1^2 \|x_1\|^2}{\sigma^2}) < \alpha \dfrac{\chi^2(n-2,0)}{n-2}\}$ so $p_2 \to 0$. Asymptotically we

will be choosing only from the two true models. The probability of picking the simplest true model, p_1, is given by

$$Pr\{e_1'e_1 + \alpha s^2 < y'y, \ e_1'e_1 < e_2'e_2, \ e_1'e_1 < e'e + \alpha s^2\} \qquad (6)$$

It is easily shown that the first two conditions will be satisfied jointly with a probability tending to one when $n \to \infty$. So, since $e_1'e_1 - e'e = b_2^2 \|x_{2.1}\|^2$

$$p_1 \to \lim_{n \to \infty} Pr\{\dfrac{b_2^2 \|x_{2.1}\|^2}{s^2} < \alpha\} = Pr\{\chi^2(1,0) < \alpha\} \qquad (7)$$

and of course $p_{12} \to Pr\{\chi^2(1,0) > \alpha\}$. To summarize

$$(p_0, p_1, p_2, p_{12}) \rightarrow (0, Pr\{\chi^2(1,0) < \alpha\}, 0, Pr\{\chi^2(1,0) > \alpha\}) \tag{8}$$

a well known result. A somewhat more detailed investigation easily reveals that the agreement between exact and asymptotic probabilities is enhanced by increasing values of n and $\beta_1^2 \|x_1\|^2 / \sigma^2$ and by decreasing values of r^2 and α.

We have performed a few experiments to test the usefulness of (8). The exact probabilities can be expressed in terms of 3 independent random variables, two standard normal variables and one $\chi^2(n-2,0)$-variable, plus of course n, $\beta_1^2 \|x_1\|^2 / \sigma^2$, α and r^2. By averaging over 10.000 independent drawings from $N(0,1) \times N(0,1) \times \chi^2(n-2,0)$ we obtained the following results as displayed in table 1, see the next page. The parameter values are chosen such that for $\sigma^2 = 1$ the average squared multiple correlation coefficient between y and X is about .5, for $\sigma^2 = 1/9$ it is about .9. Table 1 is to be compared with table 2. The results are intuitively acceptable.

Incidentally, we have also performed some experiments for the case where both β_1 and β_2 differ from zero. It appears that (p_0, p_1, p_2, p_{12}) quickly moves away from its value for $\beta_2 = 0$ to (0,0,0,1), its asymptotic value. This fact will be relevant for the bootstrap (section 4).

3. SAMPLING DISTRIBUTION

In this section we will study some aspects of the distribution of the coefficient estimators as induced by the data-based model choice. Let I_0, I_1, I_2 and I_{12} be the indicators of the subsets of the y-space corresponding with the respective choices. Then the estimator $\hat{\beta}$ is defined by

$$\hat{\beta} \equiv 0.I_0 + \begin{bmatrix} b_1. \\ 0 \end{bmatrix} .I_1 + \begin{bmatrix} 0 \\ b_{.2} \end{bmatrix} .I_2 + \begin{bmatrix} b_1 \\ b_2 \end{bmatrix} .I_{12} \tag{9}$$

$$= \begin{bmatrix} b_1. I_1 + b_1 I_{12} \\ b_{.2} I_2 + b_2 I_{12} \end{bmatrix}$$

We will focus on the standardized estimation error which by definition equals

Table 1. Estimated selection probabilities (in percentages) based on 10.000 trials; $\beta' = [1,0]$, $\|x_1\|^2 = n$.

1a. $\sigma^2 = 1$, $n = 20$

	r = 0				r = cos 45 = .71				r = cos 15 = .97			
	P_0	P_1	P_2	P_{12}	P_0	P_1	P_2	P_{12}	P_0	P_1	P_2	P_{12}
$\alpha = 1$	0	67	0	33	0	66	1	33	0	55	27	18
2	0	83	0	17	0	81	3	16	0	63	28	9
log 20	0	90	0	10	0	88	4	8	0	66	28	5

1b. $\sigma^2 = 1$, $n = 40$

	r = 0				r = .71				r = .97			
	P_0	P_1	P_2	P_{12}	P_0	P_1	P_2	P_{12}	P_0	P_1	P_2	P_{12}
$\alpha = 1$	0	67	0	33	0	68	0	32	0	63	20	17
2	0	83	0	17	0	84	0	16	0	71	21	8
log 40	0	94	0	6	0	94	0	6	0	76	21	3

1c. $\sigma^2 = \frac{1}{9}$, $n = 20$

	r = 0				r = .71				r = .97			
	P_0	P_1	P_2	P_{12}	P_0	P_1	P_2	P_{12}	P_0	P_1	P_2	P_{12}
$\alpha = 1$	0	67	0	33	0	67	0	33	0	67	1	32
2	0	82	0	18	0	82	0	18	0	82	2	16
log 20	0	89	0	11	0	90	0	10	0	89	3	8

1d. $\sigma^2 = \frac{1}{9}$, $n = 40$

	r = 0				r = .71				r = .97			
	P_0	P_1	P_2	P_{12}	P_0	P_1	P_2	P_{12}	P_0	P_1	P_2	P_{12}
$\alpha = 1$	0	68	0	32	0	69	0	31	0	67	0	33
2	0	84	0	16	0	84	0	16	0	83	0	17
$\alpha = $ log 40	0	94	0	6	0	94	0	6	0	94	0	6

Table 2. Asymptotic selection probabilities

	P_0	P_1	P_2	P_{12}
$\alpha = 1$	0	68	0	32
$\alpha = 2$	0	84	0	16
$\alpha = $ log 20	0	92	0	8
$\alpha = $ log 40	0	95	0	5

$$
\begin{bmatrix}
\dfrac{\|x_1\|}{\sigma}(\hat{\beta}_1 - \beta_1) \\[2ex]
\dfrac{\|x_2\|}{\sigma}\hat{\beta}_2
\end{bmatrix}
\tag{10}
$$

The distribution can be expressed in terms of 3 independent random variables, two $N(0,1)$-variables and one $\chi^2(n-2,0)$-variable, plus $\beta_1\|x_1\|/\sigma$, r, n and α.

However, its exact form is difficult to ascertain. We have therefore resorted to simulation and to the use of asymptotic theory. The following sub-sections collect some of the information we have about joint, marginal and conditional distributions (to save space, the proofs will be skipped).

3.1. Unconditional distributions

Section 2 suggests that in sufficiently large samples $\hat{\beta}'$ is equal to $(b_{1.},0)$ when $b_2^2\|x_{2.1}\|^2/s^2$ is less than α and $\hat{\beta}' = (b_1,b_2)$ otherwise. Therefore, it will not come as a surprise that the standardized estimation error can be shown to be asymptotically distributed as the vector v with

$$
v = \begin{bmatrix} z_1 \\ 0 \end{bmatrix} \text{ when } \frac{z_2 - r z_1}{\sqrt{(1-r^2)}}^2 < \alpha \text{ and } v = \begin{bmatrix} \dfrac{z_1 - r z_2}{1-r^2} \\[2ex] \dfrac{z_2 - r z_1}{1-r^2} \end{bmatrix} \text{ otherwise} \tag{11}
$$

where $z \sim N_2(0, \begin{bmatrix} 1 & r \\ & 1 \end{bmatrix})$, cf. (3).

The density of v_1 can be written as

$$
f_1(v_1) \equiv \phi(v_1).\Pr\{\chi^2(1,0) < \alpha\} + \sqrt{(1-r^2)}\phi(v_1\sqrt{(1-r^2)})\times
$$

$$
\times \Pr\{\chi^2(1,r^2v_1^2) > \frac{\alpha}{1-r^2}\} \tag{12}
$$

where $\phi(.)$ is the standard-normal density. The function $f_1(.)$ is a combination of the densities of z_1 and $(z_1 - r z_2)/(1-r^2)$. It is symmetrical around zero, so $\hat{\beta}_1$ is "asymptotically unbiased". For $r^2 \neq 0$ it is tri-modal, the highest mode occurring at zero and there are two small humps at the tails. Far out in the tails $f_1(v_1)$ is virtually equal to $\sqrt{(1-r^2)}\phi(v_1\sqrt{(1-r^2)})$. The variance of v_1 can be written as

$$\frac{1 - r^2 \Pr\{\chi^2(3,0) < \alpha\}}{1-r^2} \tag{13}$$

The density of v_2, $f_2(.)$, can be obtained by assigning the mass of the distribution of $(z_2-rz_1)/(1-r^2)$ on the interval $(-\sqrt{(\alpha/(1-r^2))}, +\sqrt{(\alpha/(1-r^2))})$ to zero; outside of this interval v_2 and $(z_2-rz_1)/(1-r^2)$ have the same density. So $f_2(.)$ is also symmetrical around zero. The variance of v_2 is equal to

$$\frac{\Pr\{\chi^2(3,0) > \alpha\}}{1-r^2} \tag{14}$$

The correlation between v_1 and v_2 equals

$$-r \times \frac{\Pr(\chi^2(3,0) > \alpha\}}{1-r^2 \Pr\{\chi^2(3,0) < \alpha\}} \tag{15}$$

whose square is less than r^2 (the squared correlation between b_1 and b_2).

It is interesting to compare the asymptotic covariance matrices of $[\|x_1\|(b_1-\beta_1)/\sigma, \|x_2\|b_2/\sigma]'$ and of (10). The p.s.d. difference equals

$$\frac{\Pr\{\chi^2(3,0) < \alpha\}}{1-r^2} \begin{bmatrix} r^2 & -r \\ -r & 1 \end{bmatrix} \tag{16}$$

Substitution of $\Pr\{\chi^2(3,0) > \alpha\}$ for $\Pr\{\chi^2(3,0) < \alpha\}$ yields the p.s.d. difference between the asymptotic covariance matrices of (10) and of $[\|x_1\|(b_{1.}-\beta_1)/\sigma,0]'$.

A useful scalar measure of the performance of estimators can be defined as follows. Suppose that for $\tilde{\beta}$, say, $\|X(\tilde{\beta}-\beta)\|^2/\sigma^2$ tends in distribution to a random variable u, say. Suppose also that Eu exists. Then call this number the (asymptotic) mean squared error of $\tilde{\beta}$, and denote it by $MSE(\tilde{\beta})$. Clearly,

$$MSE(b) = 2$$

$$MSE(b_{1.},0) = 1$$

It can be shown that

$$MSE(\hat{\beta}) = 1 + Pr\{\chi^2(3,0) > \alpha\} \tag{17}$$

For α = 1, 2, log 20 and log 40 the MSE equals 1.80, 1.57, 1.39 and 1.30 resp. So a model search is efficient relative to always using the full model and inefficient relative to always using the simplest true model. (We note in passing that some of the results obtained can be and are in fact generalized to less trivial situations, see e.g. Judge & Bock (1978), Dijkstra (1983) and the references therein to the pre-test literature).

Table 3 on the next page contains some information about the exact distribution of $\hat{\beta}-\beta$ obtained through 10.000 monte carlo trials, see the columns headed by "mc". We have also calculated the values based on asymptotic theory, they are to be found in the columns under "a". The parameter variation is the same as in section 2. We took for X'X:

$$n \begin{bmatrix} 1 & r \\ r & 1 \end{bmatrix} \tag{18}$$

The abbreviations "se" and "corr" stand for "standard error" and "correlation" resp.; "ub" is the upper bound of the symmetrial 90%-interval around zero. Except for o^2 = 1 and r = .97 the simulation results are reasonably well predicted by asymptotic theory.
Similar tables are available for α = 1, log 20 and log 40. The case presented, α = 2, occupies in many ways an intermediate position. The agreement between monte carlo and asymptotic results is better for α = 1 and worse for α = log n. This applies to the shape of the distributions as well as to the calculated characteristics. The bias is in absolute value an increasing function of α, the variance is decreasing in α.

3.2. Conditional distributions

Suppose a model search has led to model (1,2), i.e. both columns of X appear to carry Ey. In this case β is estimated by $\hat{\beta}$ = b, whose distribution given that the sample favors model (1,2) will in general be different from b's unconditional distribution. Similar statements are true for $\hat{\beta}|1$ and $\hat{\beta}|2$.

We will discuss $\hat{\beta}|1$ first. Clearly $\hat{\beta}|1 \sim (b_1, 0)'|1$. For n sufficiently large, all samples favoring model 1 are characterized by

Table 3. Estimated sampling characteristics of $\hat{\beta}-\beta$ for α = 2; β' = [1,0] The number of trials is 10.000.

3a. $\sigma^2 = 1$, n = 20

	$\hat{\beta}_1-\beta_1$ r = 0 mc	a	r = .71 mc	a	r = .97 mc	a	$\hat{\beta}_2$ r = 0 mc	a	r = .71 mc	a	r = .97 mc	a	$(\hat{\beta}_1,\hat{\beta}_2)$ r = 0 mc	a	r = .71 mc	a	r = .97 mc	a
mean	.00	0	-.01	0	-.16	0	.00	0	.01	0	.16	0	corr .00	0	-.67	-.60	-.95	-.94
se	.22	.22	.31	.28	.72	.67	.17	.17	.27	.24	.72	.65						
ub	.37	.37	.48	-	1.00	-	.37	.37	.54	.52	1.28	1.42						

3b. $\sigma^2 = 1$, n = 40

	$\hat{\beta}_1-\beta_1$ r = 0 mc	a	r = .71 mc	a	r = .97 mc	a	$\hat{\beta}_2$ r = 0 mc	a	r = .71 mc	a	r = .97 mc	a	$(\hat{\beta}_1,\hat{\beta}_2)$ r = 0 mc	a	r = .71 mc	a	r = .97 mc	a
mean	.00	0	.00	0	-.11	0	.00	0	.00	0	.12	0	corr .01	0	-.61	-.60	-.96	-.94
se	.16	.16	.20	.20	.56	.47	.12	.12	.17	.17	.55	.46						
ub	.26	.26	.33	-	1.00*	-	.26	.26	.37	.37	1.07	1.00						

*the associated confidence level is larger than 90% due to the peculiar shape of the distribution.

3c. $\sigma^2 = \frac{1}{9}$, n = 20

	$\hat{\beta}_1-\beta_1$ r = 0 mc	a	r = .71 mc	a	r = .97 mc	a	$\hat{\beta}_2$ r = 0 mc	a	r = .71 mc	a	r = .97 mc	a	$(\hat{\beta}_1,\hat{\beta}_2)$ r = 0 mc	a	r = .71 mc	a	r = .97 mc	a
mean	.00	0	.00	0	.00	0	.00	0	.00	0	.00	0	corr .00	0	-.61	-.60	-.95	-.94
se	.07	.07	.09	.09	.25	.22	.06	.06	.08	.08	.24	.22						
ub	.12	.12	.15	-	.46	-	.12	.12	.17	.17	.48	.47						

3d. $\sigma^2 = \frac{1}{9}$, n = 40

	$\hat{\beta}_1-\beta_1$ r = 0 mc	a	r = .71 mc	a	r = .97 mc	a	$\hat{\beta}_2$ r = 0 mc	a	r = .71 mc	a	r = .97 mc	a	$(\hat{\beta}_1,\hat{\beta}_2)$ r = 0 mc	a	r = .71 mc	a	r = .97 mc	a
mean	.00	0	.00	0	.00	0	.00	0	.00	0	.00	0	corr .02	0	-.60	-.60	-.94	-.94
se	.05	.05	.07	.07	.16	.16	.04	.04	.06	.06	.15	.15						
ub	.09	.09	.11	-	.33	-	.09	.09	.12	.12	.33	.34						

the property that $\{b_2^2 \|x_{2.1}\|^2/s^2 < \alpha\}$. Note that $b_{1.}$, b_2 and s^2 are stochastically independent. So, asymptotically, $\hat{\beta}|1 \sim (b_{1.},0)'$. More precisely, we can prove that

$$\frac{\|x_1\|}{\sigma} (\hat{\beta}_1 - \beta_1)|1 \overset{L}{\to} z_1 \tag{19}$$

So, <u>if</u> we have hit upon the simplest true model we can use the <u>uncondi</u>tional distribution to evaluate the accuracy of the ensuing estimator, provided the sample is sufficiently large. The latter proviso does not appear to be a real restriction. We have studied the distribution of $(\hat{\beta}_1 - \beta_1)$ for those monte carlo trials of section 3.1 leading to model 1, for all situations described there plus $\alpha = 2$ and $\alpha = \log n$. The smallest number of trials was about 5500. The plotted sampling distribution were perfectly bell-shaped, the probability value of the $\chi^2(2,0)$-statistic based on skewness and kurtosis averaged about 55%. The first and second order moments, the upper bounds of the 90% confidence intervals, and the MSE's were accurately predicted by asymptotic theory.

It is not so easy to analyse $\hat{\beta}|2$. In fact, we have little to offer except for the following observation. For the six situations studied in which at least 2000 trials led to model 2 (namely: $\sigma^2 = 1$; $r = \cos 15$; $n = 20,40$; $\alpha = 1,2,\log n$), we found that the distribution of $\|x_2\|(b_{.2} - r\beta_1\|x_1\|/\|x_2\|)/\sigma$ is well described by a standard normal distribution except for a slight positive bias (about +3%). The conditional MSE's were 2.3 for $n = 20$ and 3.6 for $n = 40$.

The remainder of this section will be devoted to $\hat{\beta}|(1,2)$. It can be shown that, asymptotically, $\hat{\beta}|(1,2)$ is distributed as $(b_1,b_2)' |\{\|b_2 x_{2.1}\|^2/s^2 > \alpha\}$, more precisely

$$\begin{bmatrix} \dfrac{\|x_1\|}{\sigma} (\hat{\beta}_1 - \beta_1) \\[2ex] \dfrac{\|x_2\|}{\sigma} \hat{\beta}_2 \end{bmatrix} \Bigg|(1,2) \overset{L}{\to} \begin{bmatrix} \dfrac{z_1 - rz_2}{1-r^2} \\[2ex] \dfrac{z_2 - rz_1}{1-r^2} \end{bmatrix} \Bigg|\left\{\left[\dfrac{z_2 - rz_1}{\sqrt{(1-r^2)}}\right]^2 > \alpha\right\} \tag{20}$$

The conditional density $g_1(.)$ of the first component of the righthand-side of (20) equals:

$$g_1(w_1) = \sqrt{(1-r^2)}\phi(w_1\sqrt{(1-r^2)}) \times \frac{Pr\{\chi^2(1,r^2w_1^2) > \alpha/(1-r^2)\}}{Pr\{\chi^2(1,0) > \alpha\}} \qquad (21)$$

$g_1(.)$ is symmetrical around zero, it is bi-modal with the modes further apart for larger values of r^2 (the probability ratio is less than 1 for small w_1^2 and larger than one for large w_1^2). The variance is given by

$$\frac{1 - r^2(1-\gamma)}{1-r^2} \qquad (22)$$

where $\gamma \equiv Pr\{\chi^2(3,0) > \alpha\}/Pr\{\chi^2(1,0) > \alpha\}$.

The conditional density $g_2(.)$ of the second component is easily seen to be

$$g_2(w_2) = \begin{cases} 0 & w_2^2 < \alpha/(1-r^2) \\ \dfrac{\sqrt{(1-r^2)}\phi(w_2\sqrt{(1-r^2)})}{Pr\{\chi^2(1,0) > \alpha\}} & \text{elsewhere} \end{cases} \qquad (23)$$

with variance

$$\frac{\gamma}{1-r^2} \qquad (24)$$

The conditional correlation equals

$$-r \cdot \frac{\gamma}{1+r^2(\gamma-1)} \qquad (25)$$

whose square is larger than r^2.

Finally, it is interesting to note that the conditional mean squared error $MSE(b_1,b_2|1,2)$, is larger than 2, the unconditional mean squared error, because

$$MSE(b_1,b_2|1,2) = 1 + \frac{Pr\{\chi^2(3,0) > \alpha\}}{Pr\{\chi^2(1,0) > \alpha\}} \qquad (26)$$

For $\alpha = 1, 2, \log 20$ and $\log 40$ the MSE equals 3.53, 4.64, 5.70 and 6.42 resp.

Table 4 on the next page collects some of the information on $(\hat{\beta}-\beta)|(1,2)$ obtained on the basis of those monte carlo trials leading to model

Table 4. Estimated sampling characteristics of $(\hat{\beta}-\beta)|(1,2)$ for $\alpha = 2$; $\beta' = [1,0]$.

$(\hat{\beta}_1-\beta_1)|(1,2)$

	r = 0		r = .71		r = .97	
	mc	a	mc	a	mc	a
4a. $\sigma^2 = 1$, n = 20						
mean	.01	0	.08	0	1.24	0
se	.22	.22	.42	.48	1.03	1.61
ub	.37	.37	.64	–	2.24	–
4b. $\sigma^2 = 1$, n = 40						
mean	.01	0	.00	0	1.04	0
se	.15	.16	.33	.34	.44	1.14
ub	.25	.26	.50	–	1.49	–
4c. $\sigma^2 = \frac{1}{9}$, n = 20						
mean	.00	0	.00	0	.09	0
se	.08	.07	.15	.16	.48	.54
ub	.13	.12	.24	–	.63	–
4d. $\sigma^2 = \frac{1}{9}$, n = 40						
mean	.00	0	.00	0	.01	0
se	.05	.05	.11	.11	.37	.38
ub	.09	.09	.17	–	.49	–

$(\hat{\beta}_2)|(1,2)$

	r = 0		r = .71		r = .97	
	mc	a	mc	a	mc	a
4a.						
mean	.01	0	-.06	0	-1.30	0
se	.41	.43	.56	.60	1.01	1.65
ub	.54	.54	.74	.76	2.20	2.09
4b.						
mean	-.01	0	.00	0	-1.08	0
se	.30	.30	.41	.43	.41	1.17
ub	.38	.38	.52	.54	1.50	1.48
4c.						
mean	.00	0	.00	0	-.09	0
se	.14	.14	.19	.20	.50	.55
ub	.18	.18	.25	.25	.64	.70
4d.						
mean	.00	0	.00	0	.00	0
se	.10	.10	.14	.14	.38	.39
ub	.13	.13	.18	.18	.48	.49

$(\hat{\beta}_1,\hat{\beta}_2)|(1,2)$

	r = 0		r = .71		r = .97	
	mc	a	mc	a	mc	a
4a.						
corr	.01	0	-.87	-.89	-.98	-.99
#	1740		1621		920	
4b.						
corr	.04	0	-.89	-.89	-.93	-.99
#	1687		1645		817	
4c.						
corr	.02	0	-.87	-.89	-.99	-.99
#	1778		1787		1596	
4d.						
corr	.04	0	-.88	-.89	-.99	-.99
#	1646		1607		1661	

(1,2). The precise number of available trials is indicated by #. The structure of the table and the meaning of the abbreviations are the same as for table 3. Similar tables are available for α = 1, log 20 and log 40. Again, the case presented occupies an intermediate position. The agreement between monte carlo and asymptotic results is better for α = 1 and worse for α = log n. The conditional absolute bias and the conditional variance are increasing functions of α. It is interesting to note that the variances can be dramatically overestimated by asymptotic theory, especially for r = .97, in contrast with the unconditional case where the opposite happens. Also note that the conditional bias can be excessively large.

4. SELECTION PROBABILITIES, UNCONDITIONAL DISTRIBUTIONS AND THE BOOTSTRAP

A number of authors, among them Efron (1982) and Verbeek (1984), have suggested to use the bootstrap in situations in which the model is selected and evaluated on the basis of the same data. I.e. one first estimates a model which is general enough to encompass all models to be considered. Then one simulates the model selection and the parameter estimation using the empirical distribution of the estimated residuals. Verbeek (1984, p. 27) expects this to produce meaningful results if the sample is sufficiently large and the properties to be estimated do not depend too strongly on the true data generating mechanism. We will return to the latter proviso below. First we will briefly indicate how we implemented the bootstrap.

Thousand independent y-vectors of order n×1 were generated through $y = 1 \cdot x_1 + \epsilon$ with $\epsilon \sim N_n(0, \sigma^2 I)$; $x_1 \equiv \tilde{x}_1$ and $x_2 \equiv r\tilde{x}_1 + \sqrt{(1-r^2)}\tilde{x}_2$ where $\tilde{x}_1$ and $\tilde{x}_2$ contain the values of two orthogonal polynomials of degree 1 and 2 resp., with zero mean and sums of squares equal to n. So

$$X'X = n \begin{bmatrix} 1 & r \\ & 1 \end{bmatrix}$$

No other X-columns were considered. I.e. the full model encompassing all alternatives was taken to be $y = \beta_1 x_1 + \beta_2 x_2 + \epsilon$. For each trial b_1, b_2 and e were calculated; the elements of e centered around their mean value and inflated in the usual way. The model selection and parameter estimation were then simulated using 400 independent samples of size n from the empirical distribution of the elements of e.

So, for given y, each of the 400 replications yielded a choice from the four alternatives considered. The relative frequences with which the choices were made are estimates of p_0, p_1, p_2, p_{12}. Thousand trials yield an estimate of their sampling distribution. Table 5 on the next page displays the mean values. On comparing with table 1 it is evident that the bootstrap is heavily biased at the expense of p_1 (median values are more favorable but a substantial discrepancy remains). This phenomenon can be explained if we assume that sampling from e with the corresponding b-vector fixed, i.e. bootstrapping, will yield results comparable with sampling from $N_n(0,o^2I)$ with $\beta = b$. The experiments we performed indicate that the selection probabilities are quite sensitive to small changes in the position of Ey relative to the x_1-axis, so the proviso mentioned above is not satisfied. In fact, when Ey shifts away from the x_1-axis p_1 quickly decreases, p_{12} increases and p_2 increases or decreases (if possible at all) depending on whether Ey moves towards or away from the x_2-axis (p_0 remained negligible). Now in none of the trials Xb will lie along the x_1-axis, so the bootstrap will typically produce estimates of selection probabilities "more like" (0,0,0,1) than like the true values. Consequently, averaging must yield results like those in table 5. Freedman, these proceedings, see in particular the discussion around formula (30), noted that the average value of p_1 as obtained by the bootstrap is approximately equal to $Pr\{|z+z'| < \sqrt{\alpha}\}$ where z and z' are independent standard normals. For $\alpha = 1,2$, log 20 and log 40 we get 52, 68, 78 and 83% resp. We expect to obtain better estimates of selection probabilities in a simulation study in which the bootstrap is only performed for those y-vectors favoring the simplest true model, and bootstrap samples are constructed from the elements of e_1. We have also plotted the marginal distributions of the bootstrap estimates. Broadly speaking, the distributions for p_1 and p_{12} are very skew with a heavy tail to the left and to the right resp.; for p_0 and p_2 one finds "L-shaped" distributions.

In general, we do not believe that bootstrapping with the general model will yield reliable estimates of selection probabilities.

In a similar way, for each y we got an estimate of the sampling distribution of $\hat{\beta}-\beta$, i.e. we obtained 400 values of $\hat{\beta}^*$, say, minus b. It

Table 5. Average values, based on 1000 trials, of selection probabilities estimated by the bootstrap with 400 replications;
$\beta' = [1,0]$.

5a. $\sigma^2 = 1$, n = 20

	r = 0				r = .71				r = .97			
	p_0	p_1	p_2	p_{12}	p_0	p_1	p_2	p_{12}	p_0	p_1	p_2	p_{12}
$\alpha = 1$	0	50	0	50	0	50	4	46	0	42	25	33
2	1	65	0	34	1	64	6	29	1	49	28	22
log 20	2	73	1	24	2	71	7	20	2	54	29	15

5b. $\sigma^2 = 1$, n = 40

	r = 0				r = .71				r = .97			
	p_0	p_1	p_2	p_{12}	p_0	p_1	p_2	p_{12}	p_0	p_1	p_2	p_{12}
$\alpha = 1$	0	51	0	49	0	51	1	49	0	46	25	29
2	0	67	0	33	0	67	1	32	0	55	27	18
log 40	0	81	0	19	0	80	2	17	0	62	28	10

5c. $\sigma^2 = \frac{1}{9}$, n = 20

	r = 0				r = .71				r = .97			
	p_0	p_1	p_2	p_{12}	p_0	p_1	p_2	p_{12}	p_0	p_1	p_2	p_{12}
$\alpha = 1$	0	50	0	50	0	50	0	50	0	50	3	46
2	0	66	0	34	0	66	0	34	0	66	6	29
log 20	0	75	0	25	0	75	0	25	0	74	7	19

5d. $\sigma^2 = \frac{1}{9}$, n = 40

	r = 0				r = .71				r = .97			
	p_0	p_1	p_2	p_{12}	p_0	p_1	p_2	p_{12}	p_0	p_1	p_2	p_{12}
$\alpha = 1$	0	51	0	49	0	51	0	49	0	51	0	49
2	0	67	0	33	0	67	0	33	0	67	1	32
$\alpha = $ log 40	0	81	0	19	0	81	0	19	0	81	2	17

is a bit of a puzzle how to present one thousand bivariate distributions.
A possibility is to calculate characteristics like mean values, standard
errors, correlations etc., to plot their distributions and to summarize
them by their moments and other measures. We have done so for a number
of characteristics. But here we will limit ourselves to the means of
estimated expectations, standard errors, upper bounds of symmetrical
90%-intervals around zero, and correlations. Concerning the intervals,
we have also noted what percentage of the 1000 trials produced intervals
containing the true parameter values: the confidence level (cl). Table 6,
see below, gives the information for $\alpha = 2$, where for ease of comparison
part of table 4 is repeated (the columns headed by "mc"). Similar tables
are constructed for $\alpha = 1$ and $\alpha = \log n$. There appear to be a few
patterns to discern. On the average, the absolute bias is underestimated,
the standard errors and the correlation are overestimated. (The medians
offer if anything a slightly improved picture). Perhaps this can also be
attributed to the fact that the bootstrap-samples favor the full model
disproportionately: the bootstrap tends to mimic the sampling distribution of b. This distribution has the correct means and the variances
are larger than those of $\hat{\beta}_i - \beta_i$; in addition b_1 and b_2 are more strongly
correlated than $\hat{\beta}_1$ and $\hat{\beta}_2$.

As with the selection probabilities one might expect a better
performance when in case x_1 is selected the bootstrap samples are
constructed from e_1.

5. DATA MINING AND THE ASSESSMENT OF COEFFICIENT ESTIMATES

Now we will discuss some aspects of the issue mentioned in the
introduction: how to assess the reliability of coefficient estimates
when a model search has taken place. We will focus mainly on a procedure
popular in practice, which is simply to ignore the fact that the data
are used twice, once for model selection and once for model evaluation.

The rationale behind this procedure may be something like the
following: Suppose that, through minimization of (1) over the set of
relevant models, we selected the columns $j_1, j_2, \ldots, j_q$. We may assume
that they are the first q columns of X, and we will denote the
corresponding sub-matrix by X_+. The estimate of β is then given by

Table 6. Average values, based on 1000 trials, of bootstrap estimates (with 400 replications); α = 2; β' = [1,0].

	$\hat{\beta}_1-\beta_1$						$\hat{\beta}_2$						corr $(\hat{\beta}_1,\hat{\beta}_2)$					
	r = 0		r = .71		r = .97		r = 0		r = .71		r = .97		r = 0		r = .71		r = .97	
	mc	b	mc	b	mc	b	mc	b	mc	b	mc	b	mc	b	mc	b	mc	b
6a. $\sigma^2 = 1$, n = 20																		
mean	.00	-.01	-.02		-.16	-.10	.00	.00	.01	.02	.16	.11	.00	.00	-.67	-.72	-.95	-.96
se	.22	.31	.32		.72	.78	.17	.20	.27	.31	.72	.78						
ub	.37	.48	.54		1.00	1.19	.37	.33	.54	.49	1.28	1.21						
cl	87		91		87		98		97		84							
6b. $\sigma^2 = 1$, n = 40																		
mean	.00	.00	.00		-.11	-.08	.00	.00	.00	.00	.12	.08	.01	.00	-.61	-.70	-.96	-.97
se	.16	.20	.22		.56	.59	.12	.15	.17	.22	.55	.59						
ub	.26	.33	.37		1.00	.94	.26	.23	.37	.34	1.07	.94						
cl	88		93		83		99		99		82							
6c. $\sigma^2 = \frac{1}{9}$, n = 20																		
mean	.00	.00	.00		.00	-.02	.00	.00	.00	.00	.00	.02	.00	.00	-.61	-.69	-.95	-.97
se	.07	.09	.10		.25	.29	.06	.07	.08	.10	.24	.29						
ub	.12	.15	.16		.46	.47	.12	.11	.17	.15	.48	.47						
cl	85		92		93		98		98		97							
6d. $\sigma^2 = \frac{1}{9}$, n = 40																		
mean	.00	.00	.00		.00	.00	.00	.00	.00	.00	.00	.00	.02	.00	-.60	-.69	-.94	-.96
se	.05	.07	.07		.16	.20	.04	.05	.06	.07	.15	.20						
ub	.09	.12	.11		.33	.32	.09	.08	.12	.11	.33	.30						
cl	88		93		95		99		99		99							

$[b'_+, 0'] \equiv [y'X_+(X'_+X_+)^{-1}, 0']$ in an obvious notation. Now if Ey lies in the linear sub-space spanned by the columns of X_+, i.e. if we have selected a true model, then b_+ is a realization of a random vector with a $N_q(\beta_+, \sigma^2(X'_+X_+)^{-1})$-distribution (where of course $\beta' = [\beta'_+, 0']$). And this distribution is supposedly the relevant one for an assessment of b_+'s reliability.

However, the condition that a true model has been found is not strong enough. We must at least assume that the simplest true model is selected, so that none of the columns are redundant. Otherwise working with the hypothesis that $b_+ \sim N_q(\beta_+, \sigma^2(X'_+X_+)^{-1})$ may give an unduly optimistic picture of the reliability of b_+: the conditional distribution of b_+, given the fact that the sample favored too large a model, can show much more variability around β_+ than the unconditional distribution. To exemplify, table 7 on the next page contains for the models of the previous sections some characteristics of the distribution of $b_i - \beta_i$, $i = 1,2$. Table 7 is to be compared with table 4, in which similar information is given about $(b-\beta)$ conditional upon the choice of both x_1 and x_2. Clearly, the standard errors and the bounds of the symmetrical 90% intervals are substantially underestimated. Also recall that $MSE(b_1, b_2) = 2$, which is less than half of the conditional MSE. In addition, equation (23) suggests that the nominal significance level of say, 5%, of a test of $\beta_2 = 0$ has to be multiplied by something like $1/Pr\{\chi^2(1,0) > \alpha\}$ to get the real significance level (for $\alpha = 2$ the multiplication factor is about 6.4); in fact, when α is large enough the real level will be 100%. In short, the estimates will tend to look much better than they really are. This phenomenon is of course well known, for discussions or other examples see e.g. Mosteller and Tukey (1977), Freedman (1983), Dijkstra (in preparation).

So the use of $N_q(\beta_+, \sigma^2(X'_+X_+)^{-1})$ is not justified when the model chosen is too large. But now suppose we have picked the simplest true model, i.e. $Ey \in Im(X_+)$ but Ey does not belong to any of the linear sub-spaces generated by proper subsets of the columns of X_+. For the simple models dealt with earlier, we have seen in section 3.2 that asymptotically (and for the sample sizes considered) it is indeed justified to use the unconditional distribution. This is generally true. More precisely, we can prove that

Table 7. Standard errors and upper bounds for symmetrical 90%-
intervals around zero of the distributions of $b_i - \beta_i$.

7a. $\sigma^2 = 1$, n = 20

	r = 0	r = .71	r = .97
se	.22	.32	.87
ub	.37	.52	1.42

7b. $\sigma^2 = 1$, n = 40

	r = 0	r = .71	r = .97
se	.16	.22	.61
ub	.26	.37	1.00

7c. $\sigma^2 = \frac{1}{9}$, n = 20

	r = 0	r = .71	r = .97
se	.07	.11	.29
ub	.12	.17	.47

7d. $\sigma^2 = \frac{1}{9}$, n = 40

	r = 0	r = .71	r = .97
se	.05	.07	.20
ub	.09	.12	.34

The limiting distribution of $(X_+'X_+)^{\frac{1}{2}}(b_+ - \beta_+)$ <u>given</u> that the
q columns of X_+ are selected through minimization of (1) and
that $Ey \in Im(X_+)$ with none of the columns redundant is just
$N_q(0, \sigma^2 I)$.

To sketch the argument, recall that asymptotically the competition is
restricted to true models only. Let us temporarily change the notation:
$X_+ = X_1$, $b_+ = b_{1.}$. Let $y = X_1 b_{1.} + e_1 = X_1 b_1 + X_2 b_2 + e_{12}$ where X_2 is a
submatrix of X containing r redundant columns. Clearly $e_1 = X_{2.1} b_2 + e_{12}$
(the notation being a straightforward extension of the one given in
section 2). The event that X_1 is selected means that for all X_2 we must
have that $e_1'e_1 + \alpha q s^2 < e_{12}'e_{12} + \alpha(q+r)s^2$ or equivalently,
$b_2' X_{2.1}' X_{2.1} b_2 < \alpha r s^2$. Now $b_{1.}$ is stochastically independent of b_2 (and s^2)
and since none of the columns of X_1 are redundant, so that asymptotically
virtually every sample would select them, the conclusion follows.
Incidentally, $s_1^2 \equiv e_1'e_1/(n-q)$ is of course also conditionally consistent.

So the simple procedure of ignoring the model uncertainty when
evaluating the estimates is valid when (1) data mining has revealed the
identity of the relevant regressors and (2) the sample size is
sufficiently large. When only the second condition appears doubtful, one
might consider to use the bootstrap to estimate the conditional
distribution of b_+. Probably the best way to do this is through the
construction of bootstrap samples from the residuals of the restricted
model, not the full model (the former are more likely to lead to X_+ and,
perhaps, they are "more similar" to samples from the distribution of ε).
The first condition, that we identified the relevant regressors, is by
no means trivially satisfied. In fact, when a "large" number of
alternative models is considered it is generally not unlikely to end up
with too large a model, even asymptotically. The exemplify: the
<u>asymptotic</u> probability of selecting the smallest model from 5 nested
true models each having 1 additional parameter is just 42 and 74% resp.
for $\alpha = 1$ and $\alpha = 2$ (these numbers are easily obtained through the
recursive formulas in section 2 of Bhansali and Downham (1977)).

Perhaps the suggestion in Verbeek (1984) is useful: when the data
are used to select the model report estimates of the conditional
distribution of b_+ and of the probability of selecting X_+. Both
estimates are to be obtained with the bootstrap using the estimated
residuals of the full model. (We would suggest to use the restricted

model instead, recall section 4.)

Instead of assuming that we have made the best possible choice a more conservative approach would be to try to estimate the unconditional distribution; possibly with the bootstrap, eventhough the results of section 4 are only moderately encouraging, but a good alternative seems lacking. One might object that this procedure also presupposes the validity of a questionable assumption, namely, that the full model is true. Yes, of course, but for a model to be useful it must impose at least some unwarranted restrictions. And the risk involved in assuming the smaller model to be true seems clearly larger than for the general model. On the other hand, the gains are also larger when the former is close to correct. The pros and cons will have to be balanced and a choice will have to be made. For a more extensive discussion see Dijkstra (in preparation).

To summarize, the conditions under which the data-instigated character of a model can be ignored are rather stringent, they are certainly not automatically fulfilled. There does not appear to be an easily implementable, safe, satisfying way to evaluate models with the same data from which they arose.

REFERENCES

Atkinson, A.C. (1980): A note on the generalized information criterion for choice of a model, Biometrika, 67, 413-418.
Atkinson, A.C. (1981): Likelihood ratios, posterior odds and information criteria, Journal of Econometrics, 16, 15-20.
Bhansali, R.J. and Downham, D.Y, (1977): Some properties of the order of an autoregressive model selected by a generalization of Akaike's EPF criterion, Biometrika, 64, 547-551.
Dijkstra, T.K. (1984): Misspecification and the choice of estimators, a heuristic approach, pp. 37-55 in Dijkstra, T.K. (ed.): Misspecification Analysis, Lecture Notes in Economics and Mathematical Systems (no. 237), Springer Verlag, Berlin.
Efron, B. (1979): Bootstrap methods, another look at the jackknife, The Annals of Statistics, 7, 1-26.
Efron, B. (1982): The jackknife, the bootstrap and other resampling plans, SIAM, Philadelphia.
Freedman, D.A. (1981): Bootstrapping regression models, The Annals of Statistics, 9, 1218-1228.
Freedman, D.A. (1983): A note on screening regression equations, The American Statistician, 37, 152-155.
Geweke, J. and Meese, R. (1981): Estimating regression models of finite but unknown order, International Economic Review, 22, 55-70.
Judge, G.G. and Bock, M.E. (1978): The statistical implications of pre-test and Stein-rule estimators in econometrics, North-Holland, Amsterdam.

Leamer, E.E. (1983): Model choice and specification analysis, pp. 285-330 in Griliches, Z. and Intriligator, M.D. (eds): Handbook of Econometrics I, North-Holland, Amsterdam.

Mosteller, F. and Tukey, J.W. (1977): Data analysis and regression, Addison-Wesley, Reading, Massachusetts.

Verbeek, A. (1984): The geometry of model selection in regression, pp. 20-36 in Dijkstra, T.K. (ed.): Misspecification Analysis, Lecture Notes in Economics and Mathematical Systems (no. 237), Springer Verlag, Berlin.

AUTOCORRELATION PRE-TESTING IN LINEAR MODELS WITH AR(1) ERRORS.

Henk Folmer
Department of General Economics
Wageningen Agricultural University
Wageningen, The Netherlands

0. ABSTRACT

The first part of the paper gives an overview of the recent literature on autocorrelation pre-testing in linear models. Pre-testing is found to be preferable to pure OLS and to outperform the procedures of always correcting for autocorrelation. The Durbin pre-test estimator, applying the Durbin-Watson test of autocorrelation, compares favorably with its alternative pre-test estimators. However, especially in the case of positive autoregressive errors and a trended explanatory variable the estimated risk of the Durbin estimator increases substantially beyond values of 0.6 for the autocorrelation parameter. Moreover, the confidence intervals can be quite inaccurate.

In the second part of the paper the problem how to assess the reliability of coefficient estimates of data-instigated models is investigated. Some aspects of the distribution of the pre-test estimator (applying the Durbin-Watson test of auto-correlation and the Durbin estimator in the case the hypothesis of uncorrelated disturbances is rejected) are approximated by the bootstrap technique. The main finding is that the unbiasedness of the pre-test estimator of the regression coefficient on average also holds over the bootstrap distributions. The 90% confidence interval proportions obtained by the bootstrap technique, however, are substantially lower than in the case of the pre-test estimator. Another important result is that the Durbin estimator of the autocorrelation parameter is seriously biased downwards which has a substantial impact on the performance of the pre-test estimator.

The author wishes to thank Menno de Haan and Hans Rienks for their valuable assistance and Theo Dijkstra for his stimulating comments.

1. INTRODUCTION

Linear models with autoregressive errors have been intensively studied in econometrics. This is not surprising because, if the errors are autocorrelated, the traditional maximum likelihood least squares estimator of the regression coefficients, under the assumption of no autocorrelation, will not, in general, be efficient, although it will be unbiased. Moreover, the usual estimator of the variance will be biased and the t and F tests are no longer valid.

The usual strategy in applied research is to estimate the model by ordinary least squares (OLS), to test the estimated errors for serial independence by means of, inter alia, the Durbin-Watson statistic or the Berenblutt-Webb g-statistic and then, depending on the outcome of the test, either the OLS estimates are retained or an alternative estimator, such as the Cochrane-Orcutt two-stage procedure, the Durbin or Prais-Winsten estimator, or maximum likelihood (ML) under the assumption of autocorrelated errors, is applied to take account of the detected correlation. As pointed out by various authors, among others, Leamer (1978) and Lovell (1983) a problem inherent to this strategy of pre-testing is how to assess the reliability of coefficient estimates.

The purpose of the present paper is to study the problem of pre-testing with regard to autocorrelation in linear models. First, an overview of the recent literature on autocorrelation pre-testing is given. Next, the possibilities of approximating the distribution of the pre-test estimator of the regression coefficients and the autocorrelation parameter by the bootstrap technique will be investigated. (It should be observed that the use of the bootstrap in situations where the model is selected and evaluated on the basis of the same data set has been suggested by various authors, among others, Efron (1982) and Verbeek (1984)).

The organization of the remainder of this paper is as follows. Section 2 is devoted to the overview of the recent literature on autocorrelation pre-testing. In section 3 the design for the experiments performed in this paper are described. In section 4 the main results are presented and in section 5 some conclusions are formulated.

2. AUTOCORRELATION PRE-TESTING: SOME EVIDENCE FROM THE LITERATURE

Over the last two decades a growing interest in the implications of autocorrelation pre-testing can be noticed. The most important references in this field are Rao and Griliches (1969), Judge and Bock (1978), Nakamura and Nakamura (1978), Fombey and Guilkey (1979), King and Giles (1984) and Griffiths and Beesley (1984). From these studies, which provide Monte Carlo evidence on the performance of autocorrelation pre-test estimators, the following conclusions can be drawn.

(i)   In terms of risk (under quadratic loss), size and power (for one point of the power curve) of t-tests of the regression coefficients and mean squared error of prediction, pre-testing is preferable to OLS under the assumption of no autocorrelation (in the sequel to be denoted as "pure OLS"), in particular for values of the autocorrelation parameter $\rho > 0.2$. Moreover, pre-testing compares favorably with the procedure of always correcting for autocorrelation. More specifically, pre-testing has smaller risk for $\rho = 0$ and slightly larger risk for larger ρ.

(ii) The risk gains from autocorrelation pre-testing are (sometimes highly) dependent on the choice of estimator. Judge and Bock (1978) investigated the Cochrane-Orcutt procedure, the Prais-Winsten estimator and the Durbin estimator. The last mentioned estimator was found to be preferable. It should be observed that this estimator also gained support from the study by Rao and Griliches (1969).

Griffiths and Beesley (1984) analyzed a linear model with AR (1) or MA (1) errors[1]. The following estimators were included in their experiment: (a) the ML estimator which always assumes AR(1) errors; (b) the ML estimator which always assumes MA(1) errors; (c) the pre-test estimator which always assumes MA(1) errors; (c) the pre-test estimator which, depending on the outcome of the Durbin-Watson test, is either OLS or ML which assumes AR(1) errors (OLS-AR); and (d) the pre-test estimator which, depending on the Durbin-Watson test and the values of the likelihood function, is either OLS, ML under the assumption of AR(1) errors, or ML which assumes MA(1) errors. The OLS-AR pre-test estimator compared relatively favorably when

----

1) Both positive and negative values for ρ were investigated. As will be argued in section 3, the more interesting values are non-negative.

the explanatory variable is stationary(even under an MA error specification). For the trended explanatory variable, however, the differences in risk among the pre-test estimators mutually and relative to pure OLS and to always assuming AR(1) or MA(1) errors were rather small. Concerning the accuracy of interval estimates of the regression coefficient the OLS-AR pre-test estimator was also found to be preferable. For the trended explanatory variable, however, OLS-AR gave an overestimation of the reliability for values of $\rho \geq 0.2$, though to a less extent than the alternative estimators. For example, for $\rho = 0.4$ the proportion of "95%" confidence intervals containing the regression coefficient was 0.85 and for $\rho = 0.95$ only 0.61. This phenomenon occurred for both small (20) and large (50) numbers of observations[2].

King and Giles (1984) investigated the performances of the Durbin and the ML pre-test estimators in conjunction with the Durbin-Watson, the Berenblutt-Webb and the $s(\rho_1)$ test statistics for autocorrelation. Both in terms of risk, size, power (for one point of the power function) and mean squared error of prediction only rather small differences between the Durbin and the ML estimators were found.

The upshot of this subsection is that the Durbin and ML estimators are superior to the other estimators considered and that they perform about equally well.

(iii) The choice of significance does not affect the relative ordering of the risk functions nor does it significantly change their curvature. Concerning the magnitude, however, somewhat contradictory results are obtained for the Durbin-Watson test. Judge and Bock (1978) found that different values of α ($\alpha = 0.01$, 0.025, 0.05) do not alter the magnitudes of the risk function. King and Giles (1984), however, observed differences as large as 25% for $\alpha = 0.05$ and $\alpha = 0.5$. In this regard the following should be observed. First, the ranges over which α vary in both studies differ substantially. Secondly, Judge and Bock (1978) used the modified version of the Durbin-Watson test where the null hypothesis of no serial correlation is

----

2) It is interesting that for MA(1) models with stationary explanatory variable all estimators, including pure OLS, gave proportions remarkably close to 0.95. This phenomenon did not hold for the trended explanatory variable.

rejected if the test statistic d is smaller than the upper limit d_u for the significance level. King and Giles (1984) did not indicate how they dealt with the inconclusive region. Hence, both studies may differ with regard to the treatment of the inconclusive region which might be a possible explanation for the differences observed. Thirdly, for a large number of observations (50) the difference in estimated risk for $\alpha = 0.05$ and $\alpha = 0.5$ disappeared. Fourthly, with regard to size and power of the t-test of the regression coefficients and mean squared error of prediction the results for $\alpha = 0.05$ and $\alpha = 0.5$ were very close. Finally, for the other pre-tests investigated by King and Giles, in particular the Berenblutt-Webb test which is quite similar to the Durbin-Watson test, the differences in all respects under consideration, in particular risk, were only very slight for $\alpha = 0.05$ and $\alpha = 0.5$.

The upshot of this subsection is that for a large number of observations the choice of significance level does not lead to substantial differences in performance of the pre-test estimators with regard to the relative ordering, curvature and magnitude of the risk function, size and power of the t-test of regression coefficients and mean squared error of prediction, irrespective whether the Durbin-Watson, the Berenblutt-Webb, or the $s(\rho_1)$ tests of serial correlation are applied. For a small number of observations this probably also applies when the range over which α varies is small. If α varies over a large range differences in magnitude of the risk function may show up if the Durbin-Watson test is applied.

(iv) The performance of the estimators are moderately affected by the pre-tests applied. Judge and Bock (1978) concluded that the relative powers of the Durbin-Watson and Berenblutt-Webb tests "... are very nearly the same for the entire parameter range, as measured by the ratio of the risk functions for the pre-test estimators based on each test" (p-160). In terms of risk King and Giles (1984) found that "... the choice of pre-test seems to make very little differences when $\alpha = 0.5$. The $s(\rho_1)$ and Berenblutt-Webb pre-test estimators with $\alpha = 0.05$ often have smaller risk than the corresponding Durbin-Watson pre-test estimators when ρ is large".

Similar results are found with regard to the size and power of t-tests of regression coefficients. With respect to the mean squared error of prediction the choice of pre-test also makes little difference, although the Durbin-Watson pre-test estimators frequently have larger prediction error than the other pre-tests for $\alpha = 0.05$ and for moderate and large values of ρ.

(v) The regression matrix markedly affects the relative performance of the various strategies. This follows from both the study by King and Giles (1984) where several different design matrices reflecting a variety of economic phenomena were investigated and the study by Griffiths and Beesley (1984) where a trended and a stationary explanatory variable were explored.

3. EXPERIMENTAL DESIGN

As argued in section 2 the problems inherent to autocorrelation pre-testing are most serious in the case of trended regression models. Therefore, (the simplest version of) this kind of model will be chosen for experimentation in this analysis. The model contains only a single time-trend regressor and reads as

$$Y_t = \beta t + \epsilon_t \qquad , t = 1, 2, \ldots, T \qquad (1)$$

and

$$\epsilon_t = \rho \, \epsilon_{t-1} + \mu_t \qquad , 0 \leq \rho \leq 1 \qquad (2)$$

The true value for β will be set at 1.5. The values chosen for T are 20 and 50. Moreover, $(\mu_1, \mu_2, \ldots, \mu_T)' \sim N(0, I)$ where I is the (T x T) identity matrix and $\epsilon_0 \sim N(0, \sigma_\mu^2 / (1 - \rho^2))$ and independent of $(\mu_1, \mu_2, \ldots, \mu_T)$.

It is worth noting that the estimator properties which will be investigated (amongst others accuracy of interval estimation) do not depend on the values of β and σ_μ^2 (see Breusch, 1980). As it is difficult in practice to conceive of economic processes that generate negative AR(1) error schemes, ρ will be assigned the non-negative values 0.0, 0.5 and 0.9.

As mentioned in section 1, if $\rho = 0$ the OLS estimator of β is best linear unbiased. If $\rho > 0$, however, the OLS estimator is inefficient and the conventional tests are invalid. As ρ is unknown in practice, a pre-test of $H_0 : \rho = 0$ against the one-sided alternative $H_a : \rho > 0$ is conducted. If H_0 is not rejected the OLS estimates are retained and tests of various hypotheses about the coefficients are based on the OLS residuals. Otherwise, a modified estimator and corresponding tests are adopted. Because it is this strategy which will be investigated its experimental setting will be described in detail.

As mentioned in the preceding section, various tests of autocorrelation are available. King and Giles (1984) found that the Durbin-Watson test, which is most frequently used in practice, performed somewhat less than its alternatives. Judge and Bock (1978), who applied the modified Durbin-Watson test (applying only the upper limit of significance d_u) only observed small, if not trivial, differences. Because it performs about as well as its alternatives and especially because it is most frequently used in practice, the modified version will be adopted here. The decision rules for this test procedure amount to the following

- if $d \leq d_u$ reject H_0, and

- if $d_u < d$ do not reject H_0

where

$$d = \sum_{t=2}^{T} (e_t - e_{t-1})^2 / \sum_{t=1}^{T} e_t^2$$

with e_t the OLS residual and d_u the upper limit for the significance level of d.

As mentioned above the Durbin estimator will be applied here. This estimator is constructed in the following way. First, an estimate $\hat{\rho}$ of ρ in (2) is obtained as the coefficient of the lagged dependent variable in the model

$$y_t = \rho \, y_{t-1} + \beta t - \rho \beta \, (t-1) + \mu_t, \quad t = 2, 3, \ldots, T \qquad (3)$$

The OLS estimator of ρ in (3) is consistent and efficient (Durbin, 1960).

At the second stage, the transformation $\hat{\underline{z}} = \hat{H}\underline{y}$ and $\hat{\underline{v}} = \hat{H}\underline{t}$ are formed, where $\underline{y}$ and $\underline{t}$ are the (Tx1) vectors of observations on the dependent and the explanatory variable and $\hat{H}$ is the (TxT) matrix

$$
\hat{H} = \begin{bmatrix}
\sqrt{(1 - \hat{\rho})} & 0 & 0 & 0 & \cdots & 0 & 0 \\
-\hat{\rho} & 1 & 0 & 0 & \cdots & 0 & 0 \\
0 & -\hat{\rho} & 1 & 0 & \cdots & 0 & 0 \\
& & & & & & \\
0 & 0 & 0 & 0 & \cdots & 1 & 0 \\
0 & 0 & 0 & 0 & \cdots & -\hat{\rho} & 1
\end{bmatrix}
\tag{4}
$$

Next, simple least squares is applied to these transformed variables. The coefficient of $(t - \hat{\rho}(t-1))$ is the estimate of β.[1]

The pre-test estimator of β (denoted as $\tilde{\beta}$) can be written in the following form

$$
\tilde{\beta} = I_R(d)\hat{b} + I_A(d)b
\tag{5}
$$

where $\hat{b}$ is the Durbin estimator of β, b the OLS estimator and $I_A(d)$ and $I_R(d)$ are indicator functions with A and R the acceptance and rejection intervals, respectively, for $H_O : \rho = 0$. Similarly,

$$
\tilde{\rho} = I_R(d).\hat{\rho} + I_A(d).0
\tag{6}
$$

As mentioned above, the purpose of this paper is to investigate whether the bootstrap can be used to approximate the distribution of $(\tilde{\beta}, \tilde{\rho})$. The design for this investigation has been summarized below.

Let $\{y_1^i, y_2^i, \ldots, y_T^i\}$, $i=1, 2, \ldots, I$ be the i-th data set generated via (1), (2).

----

1) It should be observed that the Prais-Winsten transformation (4) instead of the usual Cochrane-Orcutt transformation is applied here, which improves the efficiency of the Durbin estimator.

A. Generation of the distribution of $(\tilde{\beta}, \tilde{\rho})$.

(i)   Regression of y_t^i on t gives b_i and d_i.

(ii) Apply the Durbin-Watson test. If $H_o : \rho = 0$ is not rejected the regression coefficient is b_i; otherwise calculate $\hat{\rho}_i$ as the the regression coefficient of $\hat{y}_t^i$ on $(y_{t-1}^i, \; t, \; t-1)$ and $\hat{b}_i$ as the regression coefficient of $y_{t-1}^i - \hat{\rho}_i \; y_{t-1}^i)$ on $(t - \hat{\rho}_i (t-1))$.

(iii) Repeat steps (i) and (ii) I times which gives an estimate of the distribution of $(\tilde{\beta}, \; \tilde{\rho})$.

B. Generation of the bootstrap distribution.

For i=1, 2, ..., I:

(i)   Let

$$\hat{U}_t^i = y_t^i - [\hat{\rho}_i \; y_{t-1}^i + \hat{b}_i t - \hat{\rho}_i \; \hat{b}_i (t-1)] \; , \quad t=2, \ldots, T$$

$$\bar{U}_t^i = \hat{U}_t^i - \frac{1}{T-1} \sum_{t=2}^{T} \hat{U}_t^i \qquad\qquad , \; t=2, \ldots, T$$

(ii) Construct $\hat{F}$: mass $\dfrac{1}{T-1}$ at $\bar{U}_t^i$

(iii) Construct the b-th bootstrap data set

$$y_{1,b}^i = y_1^i$$

$$y_{t,b}^i = \hat{\rho}_i \; y_{t-1,b}^i + \hat{b}_i t - \hat{\rho}_i \; \hat{b}_i (t-1) + \bar{U}_{t,b}^i$$

where $\bar{U}_{t,b}^i$ are i.i.d. from $\hat{F}$.

(iv) Apply step A (ii) to the bootstrap data set which gives $\tilde{\beta}_{i,b}$ and $\tilde{\rho}_{i,b}$.

On the basis of the bootstrap distributions the average bootstrap means $ABM(\tilde{\beta})$ and $ABM(\tilde{\rho})$ will be calculated which are defined as follows:

$$ABM(\tilde{\beta}) = \frac{1}{I} \frac{1}{B} \sum_{i} \sum_{b} \tilde{\beta}_{i,b}; \quad ABM(\tilde{\rho}) = \frac{1}{I} \frac{1}{B} \sum_{i} \sum_{b} \tilde{\rho}_{i,b}$$

Similarly, the average bootstrap variances $ABV(\tilde{\beta})$ and $ABV(\tilde{\rho})$ will be calculated:

$$ABV(\tilde{\beta}) = \frac{1}{I} \frac{1}{B} \sum_{i} \sum_{b} (\tilde{\beta}_{i,b} - \frac{1}{B} \sum_{b} \beta_{i,b})^2;$$

$$ABV(\tilde{\rho}) = \frac{1}{I} \frac{1}{B} \sum_{i} \sum_{b} (\tilde{\rho}_{i,b} - \frac{1}{B} \sum_{b} \tilde{\rho}_{i,b})^2$$

With regard to parts A and B the following aspects of the distribution of $(\tilde{\beta}, \tilde{\rho})$ and $(\tilde{\beta}^*, \tilde{\rho}^*)$ (which are the bootstrap equivalents of (5) and (6)) will be computed and compared:

(1) $\bar{\tilde{\beta}} - \beta$ and $ABM(\tilde{\beta}) - \beta$; $\bar{\tilde{\rho}} - \rho$ and $ABM(\tilde{\rho}) - \rho$;

where $\bar{\tilde{\beta}} = \dfrac{1}{I} \sum_i \tilde{\beta}_i$; $\bar{\tilde{\rho}} = \dfrac{1}{I} \sum_i \tilde{\rho}_i$.

var $\tilde{\beta}$ and $ABV(\tilde{\beta})$; var $\tilde{\rho}$ and $ABV(\tilde{\rho})$

where

var $\tilde{\beta} = \dfrac{1}{I} \sum_i (\tilde{\beta}_i - \bar{\tilde{\beta}})^2$; var $\bar{\tilde{\rho}} = \dfrac{1}{I} \sum_i (\tilde{\rho}_i - \bar{\tilde{\rho}})^2$

(2) Two-sided 90% confidence intervals for β. In the case of the pre-test estimator the intervals are of the usual form:
$$\beta_i \pm t \sqrt{\sigma^2_{\tilde{\beta}_i}} \tag{7}$$

where $\tilde{\beta}_i$ is the estimate of β obtained in the i-th trial, "t" refers to Student's t distribution and $\sigma^2_{\tilde{\beta}_i}$ is the estimate of the variance of the estimator of the regression coefficient, i.e. the estimate of the variance of the OLS estimator in the case the hypothesis of correlated disturbances is rejected and the estimate of the variance of the Durbin estimator otherwise.
In the case of the bootstrap the 90% confidence intervals will be obtained by means of the bias-corrected percentile method (Efron, 1982, pp. 82-83). It consists of taking

$$[\hat{C}^{*-1} (\Phi (2Z_0-1,64)), \hat{C}^{*-1} (\Phi (2Z_0+1,64))] \tag{8}$$

as the 90% two-sided confidence interval where Φ is the standard normal cumulative distribution function, $\hat{C}^*$ is the cumulative bootstrap distribution function and

$$Z_0 = \Phi^{-1} (\hat{C}^* (\tilde{\beta}_i)) \tag{9}$$

In addition to the fraction of estimates contained in the 90% confidence intervals, the average and variance of the lengths of the intervals will be reported.

It should finally be observed that because of the large number of regressions involved, only 400 samples for each combination of ρ and T will be generated. The number of bootstrap replications is 200.

4. RESULTS

The main results of the approximation of the distribution of the pre-test estimator by the bootstrap technique are given in Tables 1 and 2. The following features of these tables need explicit mentioning.

- The pre-test estimator of ρ increasingly underestimates ρ for increasing values of ρ. This applies to both T=20 and T=50, although to a less extent in the case of the latter. These results are in conformity with among others Griffiths and Beesley (1984). The bootstrap mimics the behaviour of the pre-test estimator although for $\rho \geq 0.5$ the bias is approximately twice as large for T=20 and four times as large for T=50. Finally, var $\tilde{\rho}$ obtained via the bootstrap technique is in most cases substantially smaller than in the case of the pre-test estimator.
- With regard to $\bar{\beta} - \beta$ the results for the pre-test estimator and the bootstrap technique are very similar. This applies to both T=20 and T=50. It should also be observed that for $\rho > 0$ var $\tilde{\beta}$ is substantially smaller for the bootstrap than for the pre-test estimator.

    The confidence interval proportions for the pre-test estimator follow the pattern of deterioration outlined in section 2. For T=50 the deterioration is less serious than for T=20. From Table 3 (where F_ρ reports the proportion of estimates of ρ in a given interval and F_β (for a given estimate of ρ) the proportion of confidence intervals containing the true value of $\beta=1.5$) it follows that the downward bias of the estimator of ρ for $\rho > 0$ is an important cause of the low pre-test confidence interval proportions. Another cause might be the variance of the Durbin estimator of ρ (cf. Griffiths and Beesley, 1984). The pattern of deterioration of confidence interval proportions in the case of the bootstrap is worse than in the case of the pre-test estimator. Moreover, compared to T=20 no improvement for T=50 can be discerned.
- On the basis of these findings the conclusion can be drawn that the application of standard theory in the case of autocorre-

Table 1

Characteristics of the distribution of the pre-est estimator (P) and of the bootstrap (*) distribution based on 400 trials, 200 bootstrap replications, T=20.

ρ	Characteristic	P	*
	$\bar{\tilde{\beta}} - \beta$ [1]	-0.0010	-0.0010
	$\bar{\tilde{\rho}} - \rho$ [1]	0.0240	0.0240
	var $\tilde{\beta}$ [2]	0.0003	0.0005
0.0	var $\tilde{\rho}$ [2]	0.0078	0.0082
	90% C.I.	90.5000	87.5000
	length C.I.	0.0660	0.0720
	var C.I.	0.0002	0.0004
	$\bar{\tilde{\beta}} - \beta$ [1]	-0.0020	-0.0020
	$\bar{\tilde{\rho}} - \rho$ [1]	-0.2440	-0.4580
	var $\tilde{\beta}$ [2]	0.0012	0.0005
0.5	var $\tilde{\rho}$ [2]	0.0490	0.0160
	90% C.I.	75.2500	56.0000
	length C.I.	0.0900	0.0720
	var C.I.	0.0011	0.0020
	$\bar{\tilde{\beta}} - \beta$ [1]	-0.0040	-0.0030
	$\bar{\tilde{\rho}} - \rho$ [1]	-0.3990	-0.7930
	var $\tilde{\beta}$ [2]	0.0120	0.0005
0.9	var $\tilde{\rho}$ [2]	0.0540	0.0300
	90% C.I.	47.2500	14.2500
	length C.I.	0.1400	0.0540
	var C.I.	0.0060	0.0025

1) ABM(.) in the case of the bootstrap.
2) ABV(.) in the case of the bootstrap.

Table 2

Characteristics of the distribution of the pre-test estimator (P) and of the bootstrap (*) distribution based on 400 trials, 200 bootstrap replications, T=50

ρ	Characteristic	P	*
	$\bar{\bar{\beta}} - \beta^{1)}$	-0.00001	-0.00001
	$\bar{\bar{\rho}} - \rho^{1)}$	0.01500	0.01800
	var $\tilde{\beta}^{2)}$	0.00002	0.00003
0.0	var $\tilde{\rho}^{2)}$	0.00350	0.00380
	90% C.I.	89.00000	85.50000
	length C.I.	0.01600	0.01800
	var C.I.	0.00010	0.00002
	$\bar{\bar{\beta}} - \beta^{1)}$	0.00100	0.00100
	$\bar{\bar{\rho}} - \rho^{1)}$	-0.09600	-0.44600
	var $\tilde{\beta}^{2)}$	0.00009	0.00004
0.5	var $\tilde{\rho}^{2)}$	0.02600	0.01000
	90% C.I.	80.50000	59.00000
	length C.I.	0.02880	0.02000
	var C.I.	0.00050	0.00007
	$\bar{\bar{\beta}} - \beta^{1)}$	0.00300	0.00200
	$\bar{\bar{\rho}} - \rho^{1)}$	-0.15100	-0.55500
	var $\tilde{\beta}^{2)}$	0.00150	0.00001
0.9	var $\tilde{\rho}^{2)}$	0.01600	0.01600
	90% C.I.	59.00000	5.75000
	length C.I.	0.07100	0.00670
	var C.I.	0.00120	0.00009

1) ABM(.) in the case of the bootstrap.
2) ABV(.) in the case of the bootstrap.

Table 3

Proportion of "90%" confidence intervals containing β=1.5 (F_β) corresponding to the proportion of estimates of ρ in a given interval (F_ρ), T=20, 400 trials

ρ	Estimation interval for ρ	F_ρ	F_β
	-1.0 - -0.001	0.00	0.00
	0	92.25	83.25
	0.0 - 0.199	1.25	1.00
0	0.2 - 0.399	5.00	4.75
	0.4 - 0.599	1.25	1.25
	0.6 - 0.799	0.25	0.25
	0.8 - 1.000	0.00	0.00
		100.00	90.50
	-1.0 - -0.001	1.50	1.00
	0	30.25	16.50
	0.0 - 0.199	9.25	5.50
0.5	0.2 - 0.399	30.00	25.25
	0.4 - 0.599	23.25	21.50
	0.6 - 0.799	5.75	5.50
	0.8 - 1.000	0.00	0.00
		100.00	75.25
	-1.0 - -0.401	0.00	0.00
	-0.4 - -0.201	0.25	0.00
	-0.2 - -0.001	1.00	0.00
	0	4.25	0.50
0.9	0.0 - 0.199	6.00	2.00
	0.2 - 0.399	18.75	5.25
	0.4 - 0.599	30.25	12.25
	0.6 - 0.794	31.75	20.75
	0.8 - 1.000	7.75	6.50
		100.00	47.25

lation pre-testing may give quite misleading results with regard to the autocorrelation parameter and the regression coefficient. In particular, 90% confidence interval proportions for the regression coefficient are likely to be much less than 0.9 for large ρ.

From the results in the Tables 1 and 2 it follows that the expectation of the pre-test estimator can be satisfactorily approximated by the bootstrap technique. (A similar result was found by Dijkstra (1987) for the selection of regressors). With regard to the variances of and 90% confidence intervals proportions for the regression coefficient for large values of ρ the bootstrap results differ substantially from the pre-test estimator results. The same holds for the expected value of the estimator of the autocorrelation parameter.

5. CONCLUSIONS

From the overview of the literature in section 2 and the simulation experiments the following conclusions can be drawn.

(i)   The regression matrix markedly affects the performance of pre-test estimators. In particular, in the case of trended explanatory variables (which has been studied in the experiments of this paper and to which the conclusions below refer) the Durbin estimator of the autocorrelation parameter is seriously biased downwards. Moreover, an increase of the risk is likely to occur. Finally, inaccurate confidence intervals may be obtained.

(ii)   Pre-testing is preferable to pure OLS for all values of ρ.

(iii) In terms of risk and interval estimation accuracy, pre-testing is preferable to always correcting for auto-correlation for small values of ρ whereas for increasing values of ρ the two strategies converge.

(iv)   The negative consequences of pre-testing such as inac-curate confidence interval and increase of the risk, are more more serious for small than for large samples.

(v)   An important cause of the negative consequences of autocorrelation pre-testing is the downward bias of the estimator of the autocorrelation parameter.

(vi)   The pre-test estimator of the regression coefficient is unbiased. Moreover, its expected value can be satisfactorily approximated by the bootstrap technique. With regard to the variance of the estimator of and confidence intervals for the regression coefficient the bootstrap and pre-test estimation results differ substantially for large ρ. The same holds for the expected value and variance of the estimator of the auto-correlation parameter.

REFERENCES

Breusch, T.S., 1980, Useful invariance results for generalized regression models, Journal of Econometrics 13, 327-340.

Dijkstra, T., 1987, Data-driven selection of regressors and the bootstrap, this volume.

Durbin, J., 1960, Estimation of parameters in time regression models, Journal of the Royal Statistical Society B 22, 139-153.

Efron, B., 1982, The Jackknife, the bootstrap and other re-sampling plans (SIAM, Philadelphia).

Fomby, T.B., and D.K. Guilkey, 1978, On choosing the optimal level of significance for the Durbin-Watson test and the Bayesian alternative, Journal of Econometrics 8, 203-213.

Griffiths, W.E. and P.A.A. Beesley, 1984, The small-sample properties of some preliminary test estimators in a linear model with autocorrelated errors, Journal of Econometrics 25, 49-61.

Judge, G.G. and M.E. Bock, 1978, The statistical implications of pre-test and Stein-rule estimators in econometrics (North-Holland, Amsterdam).

Judge, G.G., W.E. Griffiths, R.C. Hill and T.C. Lee, 1980, The theory and practice of econometrics (Wiley, New York).

King, M.L. and D.E.A. Giles, 1984, Autocorrelation pre-testing in the linear model. Estimation, testing and prediction. Journal of Econometrics 25, 35-48.

Leamer, E.E., 1978, Specification searches. Ad hoc inference with non-experimental data (Wiley, New York).

Lovell, M.C., 1983, Data mining, Review of Economics and Statistics 65, 1-12.

Nakamura, A. and M. Nakamura, 1978, On the impact of the tests for serial correlation upon the test of significance for the regression coefficient, Journal of Econometrics 7, 199-210.

Rao, P. and Z. Griliches, 1969, Small-sample properties of several two-stage regression methods in the context of auto-correlated errors, Journal of the American Statistical Association 64, 253-272.

Verbeek, A., 1984, The geometry of model selection in regression, in T.K. Dijkstra (ed), Misspecification Analysis, Lecture Notes in Economics and Mathematical analysis, no 237 (Springer, Berlin).

ON CROSS-VALIDATION FOR PREDICTOR EVALUATION IN TIME SERIES

Tom A.B. Snijders [*]
University of Groningen
Vakgroep Statistiek & Meettheorie
Groningen, The Netherlands

ABSTRACT

In the context of the prediction error method for one step ahead prediction in a single time series, a conventional and two cross-validatory procedures are proposed for prediction of squared prediction errors, and also for choosing among several predictor families. These procedures are compared in a simulation study. The conventional procedure appears to perform at least as well as the cross-validatory procedures.

1. INTRODUCTION.

Cross-validation was proposed by Stone (1974) as a predictive approach to the evaluation of performance of statistical procedures. It seems not to have been studied often in the context of time series prediction. In this paper a time series $Y_1, \ldots, Y_n$ is considered, and prediction of Y_n from $Y_1, \ldots, Y_{n-1}$ is studied; the focus is on prediction of squared prediction errors, as a performance measure of the predictor for Y_n, and on the choice among several predictor families. It is not assumed that the predictor family under consideration is "optimal" for the time series at hand. Instead, a "predictor error approach" is followed, as discussed in Ljung and Söderström (1983).

Some "conventional" and "cross-validatory" procedures are proposed, and their performance is compared in a simulation study. The cross-validatory procedures are based on realized prediction errors in the past. The conventional procedures seem to perform, on the average, at least as well as the cross-validatory procedures.

---

[*] This research was carried out when the author was at the Institute of Econometrics of the University of Groningen.

2. PREDICTION ERROR METHODS

Prediction of time series often proceeds from the assumption that the type of time series model, e.g., an ARMA model of unspecified order, is known. From the point of view of practical applications, however, it seems more sensible not to use this assumption. An attractive approach seems to be the following: for prediction of a time series $Y_1, Y_2, \ldots$ one considers a certain family of predictors

$$\hat{Y}_n = \hat{Y}_n(\theta) = f_n(Y_1, Y_2, \ldots, Y_{n-1}; \theta)$$

where $\theta \in \Theta$ for some parameter space $\Theta \subset \mathbf{R}^p$, where Θ has a non-empty interior so p is the true dimension of the predictor family, and where the f_n, for different n, have a logical consistency. Many families of predictors have a state-space form, i.e., they can be expressed with the aid of the "state variable" Z_n as

$$\hat{Y}_n = g_1(Z_n; \theta) \qquad \text{(prediction equation)}$$
$$Z_{n+1} = g_2(Z_n; Y_n; \theta) \qquad \text{(update equation)}$$
$$Z_{n(0)} = g_0(Y_1, \ldots, Y_{n(0)-1}; \theta) \qquad \text{(initial state)}$$

for $n \geq n(0)$; this is a nice way of achieving logically consistent prediction functions f_n, $n \geq n(0)$. For the given family of predictors, one tries to determine θ so as to come near to the optimal predictor within this family; for some loss function $L(Y_n, \hat{Y}_n)$ (which could also be taken to depend on θ), the optimal predictor is $\hat{Y}_n(\theta_n^*)$ where $EL(Y_n, \hat{Y}_n(\theta_n^*)) = \min_\theta EL(Y_n, \hat{Y}_n(\theta))$.

A powerful method for estimating the optimal θ is the so-called prediction error method, where θ_n^* is estimated by the value of θ, denoted $\hat{\theta}_{n-1}$, for which

$$\sum_{i=n(0)}^{n-1} L(Y_i, \hat{Y}_i(\theta)) \qquad (1)$$

is minimal. Prediction error methods are extensively treated in Ljung and Söderström (1983), where many further references are to be found. One key result proved by Ljung (1978) in a more general framework (admitting explanatory variables and control), is that under certain assumptions, including the existence of the limit

$$\liminf_{\substack{n \to \infty \\ \theta \in \Theta}} EL(Y_n, \hat{Y}_n(\theta)), \qquad (2)$$

the minimum prediction error method is consistent in the sense that (2) is equal to

$$\text{plim}_{n \to \infty} EL(Y_n, \hat{Y}_n(\hat{\theta}_{n-1})).$$

The present paper is concerned with the evaluation of a predictor family for the given time series (with unknown probability distribution), following the prediction error method with squared error loss, $L(Y_n, \hat{Y}_n) = (Y_n - \hat{Y}_n)^2$. In particular, the prediction of the squared prediction error

$$E_n^2 = (Y_n - \hat{Y}_n(\hat{\theta}_{n-1}))^2 \tag{3}$$

will be considered in Sections 3 and 4; and the data-dependent choice of predictor family will be considered in Section 5. In this context, the difference between predicting E_n^2 and estimating EE_n^2 is not essential.

3. CROSS-VALIDATION APPROACHES TO PREDICTION.

Cross-validation, as proposed by Stone (1974), seems not to have been applied often to time series problems. Shibata (1985), e.g. mentions cross-validation as an approach for model selection in time series, but does not give any reference or elaboration. Shibata's specification of cross-validation is based on deletion of single observations, unlike the approach of the present paper where all observations after a certain time point are deleted.

A self-evident approach to prediction of the squared prediction error (3) is to use a predictor based on the minimized criterion (1), with some compensation for the overfitting due to using $\hat{\theta}_{n-1}$ which minimizes (1). It seems reasonable to assume that EE_n^2 is roughly proportional to

$$1 + (n-p)^{-1} = (n-p+1)/(n-p); \tag{4}$$

recall that p is the dimensionality of the fitted parameter θ. This assumption is not compelling, however, as EE_n^2 depends both on the predictor family and the unknown time series distribution. The assumption leads to

$$V_{n1} = \frac{(n-p+1)}{(n-p)(n-n(0)-p)} \sum_{i=n(0)}^{n-1} (Y_i - \hat{Y}_i(\hat{\theta}_{n-1}))^2 \tag{5}$$

as a predictor for E_n^2. The factor $(n-p+1)/(n-p)$ in this definition follows from (4); the factor $n-n(0)-p$ in the denominator is the number of terms $n-n(0)$ minus the compensation p for overfitting in (1) by the estimation of p parameters.

A cross-validatory approach to prediction of E_n^2, using only non-interrupted time series as "validating sub-samples", is to use a predictor based on past prediction errors

$$E_{i(0)}, E_{i(0)+1}, \ldots, E_{n-1}.$$

These prediction errors $E_i = (Y_i - \hat{Y}_i(\hat{\theta}_{i-1}))$ are the errors actually realized in the past by the prediction error method. The assumption that EE_n^2 is roughly proportional to (4) leads here to

$$V_{n2} = \frac{(n-p+1)}{(n-p)(n-i(0))} \sum_{i=i(0)}^{n-1} \frac{(i-p)}{(i-p+1)} E_i^2 \qquad (6)$$

as a predictor for E_n^2. Note that earlier squared prediction errors E_i^2 are down-weighted in (6) by $(i-p)/(i-p+1)$, an increasing function of i. A pilot simulation study indicated that V_{n2} has in many cases a positive bias, and suggested to try a version of V_{n2} with a more severe downweighting factor. This led to a third predictor for E_n^2,

$$V_{n3} = \frac{(n-p+1)}{(n-p)(n-i(0))} \sum_{i=i(0)}^{n-1} \frac{(i-p-1)}{(i-p+1)} E_i^2. \qquad (7)$$

It is instructive to see the implications of these formulas for the simplest family of predictors, namely, the family of constant predictors

$$\hat{Y}_n(\theta) = \theta, \quad \theta \in R,$$

with dimension $p=1$, defined for $n \geq n(0) = 1$. The prediction error method leads here to

$$\hat{Y}_n(\hat{\theta}_{n-1}) = \hat{\theta}_{n-1} = Y.^{(n-1)} = \frac{1}{n-1} \sum_{i=1}^{n-1} Y_i .$$

Helmert's transformation (see, e.g., Kendall and Stuart, 1977) yields

$$\sum_{i=1}^{n-1} (Y_i - Y.^{(n-1)})^2 = \sum_{i=2}^{n-1} \frac{i-1}{i} (Y_i - Y.^{(i-1)})^2 .$$

This implies that, for the family of constant predictors, with $n(0)=1$ and $i(0) = 2$, one has

$$V_{n1} = V_{n2}.$$

Now consider the corresponding model that the time series $Y_1, Y_2, \ldots$ is an independent identically distributed (i.i.d.) sequence with finite variance σ^2, without any further assumptions about the distribution. Then the family of constant predictors is the most sensible choice for a predictor family. Note that the expressions given for $\hat{\theta}_{n-1}$ and V_{n1} demonstrate that these predictors are in this case symmetric functions of variables Y_1 to Y_{n-1}, i.e., they do not depend on the order of $Y_1, \ldots, Y_{n-1}$. Predicting Y_n is here equivalent to estimating EY_n, so standard results imply that $Y_.^{(n-1)}$ is the uniformly minimum variance unbiased predictor for Y_n. Similarly, predicting $E_n^2 = (Y_n - Y_.^{(n-1)})^2$ is here equivalent to estimating $n\sigma^2/(n-1)$, and under the additional assumption $EY_1^4 < \infty$ it can be concluded that V_{n1} with $n(0) = 1$ is the uniformly minimum variance unbiased predictor for E_n^2. Nothing is implied, however, for the performance of these predictors for E_n^2 when the time series does not have an i.i.d. distribution.

There is an obvious analogy of the realized prediction errors E_i $(i(0) \le i \le n-1)$ with recursive residuals introduced by Brown, Durbin and Evans (1975). Realized prediction errors can be used very well for diagnostic purposes; they reveal the instances where the considered predictor family would have failed. Also, incompatibility of the predictor family with the time series distribution will in many cases tend to show up in an increasing or other trend in the E_i^2.

4. A SIMULATION STUDY

A simulation study was carried out in order to explore the relative performance of V_{n1}, V_{n2}, and V_{n3} as predictors for E_n^2. Three predictor families were considered.

1. Constant predictors; here $p=2$ and

$$\hat{Y}_n(\theta) = \theta \qquad (n \ge 1)$$

for $\theta \in R$.

2. Linear predictors of order 1; here $p=2$ and

$$\hat{Y}_n(\theta) = \theta_1 + \theta_2 Y_{n-1} \qquad (n \ge 2)$$

for $\theta_1 \in R$, $|\theta_2| \le 1$.

3. Exponentially weighted moving average; here p=2 and

$$\hat{Y}_1(\theta) = \theta_1$$

$$\hat{Y}_n(\theta) = \theta_2\hat{Y}_{n-1}(\theta)+(1-\theta_2)Y_{n-1} \qquad (n \geq 2)$$

for $\theta_1 \in R$, $|\theta_2| \leq 1$.

The restrictions on θ_2 for predictor families 2 and 3 are needed to achieve stable predictors.

The prediction error method for the constant predictor family was worked out in the preceding section. For the family of linear predictors of order 1, a simple regression problem is obtained leading to

$$\hat{\theta}^0_{2,n-1} = \{\sum_{i=1}^{n-2} (Y_i-\bar{Y}.^{(n-2)})(Y_{i+1}-\tilde{Y}.^{(n-1)})\}/\{\sum_{i=1}^{n-2} (Y_i-\bar{Y}.^{(n-2)})^2\}$$

$$\hat{\theta}_{2,n-1} = \min \{\max \{ \hat{\theta}^0_{2,n-1},-1\} , 1\}$$

$$\hat{\theta}_{1,n-1} = \tilde{Y}.^{(n-1)}-\hat{\theta}_{2,n-1}\bar{Y}.^{(n-2)}$$

where

$$\tilde{Y}.^{(n-1)} = \frac{1}{n-2} \sum_{i=2}^{n-1} Y_i.$$

The prediction error method for the exponentially weighted moving average family must be worked out numerically. The minimization of (1) as a function of θ_1 can be carried out explicitly; a function of θ_2 results, which was minimized by a quasi-Newton method.

Each of these three predictor families was used for prediction of the following time series; in the specifications of the time series distributions, the U_i and V_i are independent standard normal random variables.

IID. A sequence of independent and identically distributed random variables:
$$Y_i = U_i \qquad\qquad i \geq 1.$$

AR1. A stationary autoregressive sequence of order 1:
$$Y_1 = (1-\eta^2)^{-1/2}U_1$$
$$Y_i = \eta Y_{i-1}+U_i \qquad\qquad i \geq 2.$$

IMA1. An integrated moving average of order 1:
$$Y_1 = 0$$
$$Y_i = Y_{i-1}+U_i+\eta U_{i-1} \qquad\qquad i \geq 2.$$

MA1. A moving average of order 1:
$$Y_i = \eta U_{i-1}+(1-\eta)U_i \qquad\qquad i \geq 1. \cdot$$

AR1+N. A stationary autoregressive sequence of order 1 plus noise:

$$Y_i = Z_i + V_i \qquad\qquad i \geq 1$$

where

$$Z_1 = (1-\eta^2)^{-1/2} U_1$$

$$Z_i = \eta Z_{i-1} + U_i \qquad\qquad i \geq 2.$$

Note that the three predictor families 1, 2 and 3 are optimal, respectively, for the IID sequence, the AR1 sequence and the IMA1 sequence; for the latter optimality see, e.g., Kendall, Stuart and Ord (1983) Chapter 47.

Prediction of the squared prediction error E_n^2 was considered for $10 \leq n \leq 15$. For each time series distribution, 1000 simulation runs were executed where in each run the three predictor families 1, 2, 3 and each of the three predictors V_{n1}, V_{n2}, V_{n3} were applied for n = 10 to 15. At the start of each simulation study the random number generator was set at a non-repeatable state based on the real time clock. Normally distributed random variables were generated by Brent's (1974) method. This was carried out using subroutines in the NAG library. The performance of each of the predictors V_{nj} was measured by the estimated mean squared prediction error, defined as the average over the 1000 simulation runs of $(E_n^2 - V_{nj})^2$.

As examples, some results for MA1 and IMA1 time series are presented in the following figures. For the constant predictor family applied to the MA1 ($\eta = .5$)

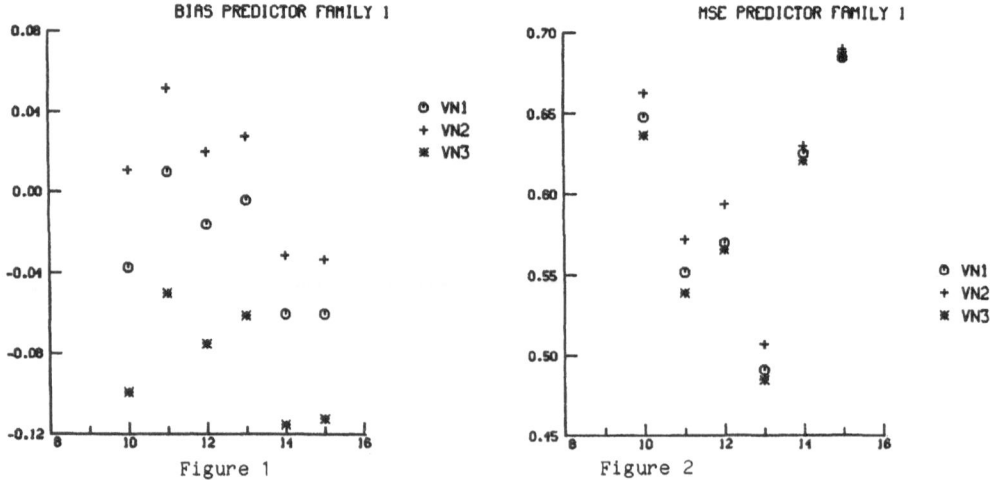

Figure 1   Figure 2

Figures 1 and 2. Estimated bias (Figure 1) and estimated mean squared prediction error (Figure 2) of V_{n1}, V_{n2} and V_{n3} for predictor family 1 with MA1 (η=.5) time series. Standard errors are about .025 for bias and about .06 for empse.

time series, V_{n3} exhibits a rather strong downward bias; V_{n1} exhibits a downward bias exceeding 2 standard errors for n = 14, 15; V_{n2} seems not very biased. As for estimated mean squared prediction errors (empse), V_{n3} tends to be best and V_{n2} tends to be worst.

For the linear predictor family applied to the same time series, the situation is different: Here the bias of V_{n1} and V_{n3} is not significantly different from 0, and V_{n2} has a considerable upward bias. The empse of V_{n1} is a bit smaller that that of V_{n3}, while the empse of V_{n2} is considerably and significantly larger. The situation is similar in most other simulations:

* V_{n1} tends to be smaller than V_{n2}; of course $V_{n3} < V_{n2}$ always;
* the bias of V_{n1} is downward or not appreciable;
* the bias of V_{n2} is upward or not appreciable;
* the mean squared prediction error of V_{n2} tends to be largest.

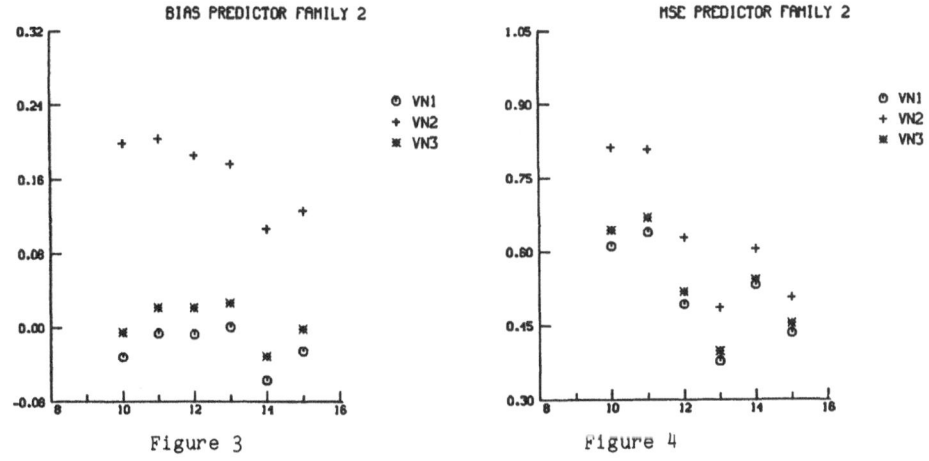

Figure 3   Figure 4

Figures 3 and 4. Estimated bias (Figure 3) and estimated mean squared prediction error (Figure 4) of V_{n1}, V_{n2} and V_{n3} for predictor family 2 with MA1 (η=.5) time series. Standard errors are about .025 for bias and about 0.06 for empse.

Only if the predictor family is incompatible with the time series distribution, a different picture is obtained. This is the case when the constant predictor family is used to predict the non-stationary IMA1 time series. Figure 5 shows that the mean of E_n^2 increases about linearly with n for this case; the limit in

(2) is infinite. The empse of V_{n1}, V_{n2} and V_{n3} also increases with n, but here V_{n1} performs significantly worse than V_{n2}. This case is not of any practical value, because a plot of past values of V_{n1} or E_n would soon reveal the inadequacy of the predictor family for the time series at hand.

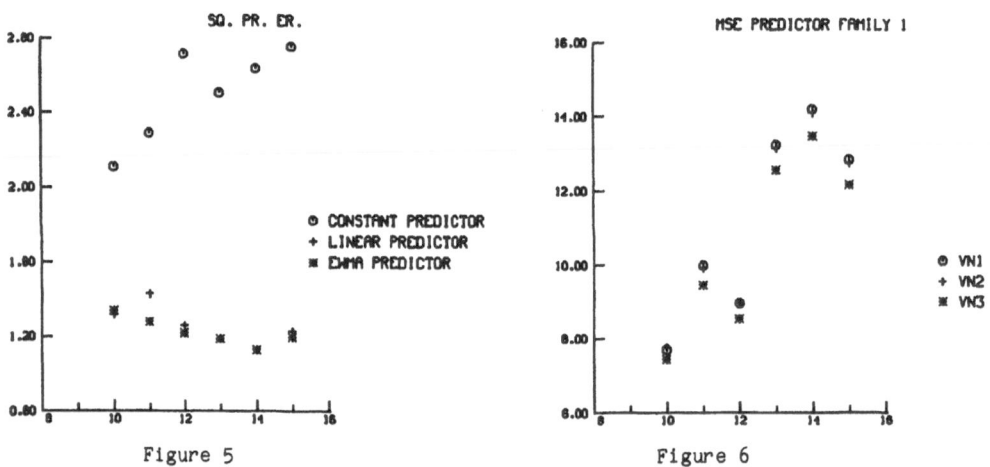

Figure 5   Figure 6

Figures 5 and 6. Estimated mean squared prediction errors of $\hat{Y}_n(\hat{\theta}_{n-1})$ (Figure 5) and estimated mean squared predictions errors of V_{n1}, V_{n2} and V_{n3} for predictor family 1 (Figure 6) with IMA1 ($\eta=-.3$) time series. Standard errors in Figure 5 range from .09 to .13 for predictor family 1, and are about .06 for predictor families 2 and 3. Standard errors in Figure 6 range from .8 to 2.2.

The results of the simulations are summarized in Table 1. The average relative efficiencies

$$\text{are}(1:j) = \underset{10 \leq n \leq 15}{\text{average}} \frac{\text{empse } (V_{nj})}{\text{empse } (V_{n1})} \qquad (8)$$

are presented for j=2,3. Predictor V_{n1} is viewed as a standard for comparison.

The standard errors of empse's are such that standard errors of are(1:j) are about 0.015. It can be concluded that on the whole, V_{n1} performs best, followed by V_{n3}. Especially for predictor family 2, V_{n2} tends to have a rather high upward bias, leading to high empse's. A small pilot study was done to investigate a predictor for E_n^2 where earlier E_i^2 values are downweighted even further than in V_{n3}, by replacing the factor i-p-1 in the formula for V_{n3} by i-p-2. This did not lead to smaller emspe's, so the investigation of this predictor was not continued.

Table 1. Average relative efficiencies (8) obtained from 1000 simulation runs.

Predictor   families

	1		2		3	
Time series	are(1:2)	are(1:3)	are(1:2)	are(1:3)	are(1:2)	are(1:3)
IID	1.03	1.01	1.16	1.03	1.09	1.03
AR1 (η=.5)	1.02	1.01	1.14	1.03	1.05	1.03
AR1 (η=.2)	1.04	1.01	1.17	1.04	1.10	1.04
IMA1 (η=-.3)	.95	1.00	1.07	1.02	1.05	1.03
IMA1 (η=.3)	.91	.97	1.16	1.05	1.05	1.02
MA1 (η=.5)	1.02	.99	1.24	1.05	1.05	1.01
MA1 (η=.2)	1.03	1.00	1.18	1.03	1.08	1.03
AR1+N (η=.5)	1.02	1.01	1.15	1.04	1.06	1.03
AR1+N (η=.2)	1.03	1.00	1.19	1.04	1.11	1.03

It can be concluded that the "conventional" predictor V_{n1} performs at least as well as the "cross-validatory" predictors V_{n2} and V_{n3}. A further discussion of the results is given in Section 5.

4. CROSS-VALIDATORY CHOICE OF PREDICTOR FAMILY

If little information is available about the type of probability distribution of the time series $Y_1, Y_2, \ldots$ then it can be difficult to choose on the basis of a **priori** considerations a suitable predictor family. This section is concerned with the choice, among a given class of predictor families, of a suitable predictor family for predicting Y_n, basing this choice on $Y_1, Y_2, \ldots, Y_{n-1}$. The only essential difference between this and the choice of an optimal predictor within a given family is constituted by possible dimensionality differences between the different families.

This paper does not attempt to discuss the selection of a true model among several competing ones, although this is obviously a related subject; a review is given by Shibata (1985). In the context of selection of a true model, much attention is given to the penalty attached to selection of a higher dimension. In the present paper, a comparison is made only between a "conventional" and a "cross-validatory" choice of predictor family.

Suppose k predictor families

$$Y_n^{(h)}(\theta), \quad \theta \epsilon \Theta^{(h)}; \quad h=1,\ldots,k$$

are available and that for each h, the value $\hat{\theta}_{n-1}^{(h)}$ produced by the prediction error method is used. A natural choice procedure for the prediction of Y_n is to choose the predictor family h for which the predicted value $V_n^{(h)}$ for the squared prediction error

$$(Y_n - \hat{Y}_n^{(h)}(\hat{\theta}_{n-1}^{(h)}))^2$$

is smallest. In this section the predictors $V_{n1}^{(h)}$, $V_{n2}^{(h)}$ and $V_{n3}^{(h)}$ defined in Section 3 are compared for choosing in this way between the families 1, 2 and 3 also studied in Section 3. The same time series distributions are considered in the simulation set-up of Section 3.

Let $\tilde{Y}_n^{(j)}$ be the predictor among $\hat{Y}_n^{(h)}(\hat{\theta}_{n-1}^{(h)})$, h=1,2,3, for which $V_{nj}^{(h)}$ is smallest. Picking the predictor family which happened to perform best (in some sense) in the past is an act which, of course, invites capitalization on chance. The "conventional" squared prediction error predictor $V_{n1}^{(h)}$ uses hindsight in a blunter way than its cross-validatory competitors $V_{n2}^{(h)}$ and $V_{n3}^{(h)}$, and therefore might be suspected to be more susceptible to chance capitalization and to perform worse. In this simulation, this suspicion did not appear to agree with numerical results. Table 2 gives the average estimated relative efficiencies

$$\text{are}(1:j) = \underset{10 \leq n \leq 15}{\text{average}} \frac{\text{empse }(\tilde{Y}_n^{(j)})}{\text{empse }(\tilde{Y}_n^{(1)})} \qquad (9)$$

for j=2,3.

Table 2. Average relative efficiencies (9) obtained from 1000 simulation runs.

Time series		are(1:2)	are(1:3)
IID		.98	.98
AR1	(η=.5)	1.01	1.00
AR1	(η=.2)	.98	.98
IMA	(η=-.3)	1.02	1.00
IMA1	(η=.3)	1.05	1.02
MA1	(η=.5)	1.01	1.00
MA1	(η=.2)	.97	.99
AR1+N	(η=.5)	.99	.99
AR1+N	(η=.2)	.99	1.00

The standard errors of empse's are such that these values are not significantly different from 1. So, at least in these cases, cross-validatory choice of predictor family performs no better and no worse than conventional choice. For each j, the minimised $V_{nj}^{(h)}$ presents an over-optimistic picture of the performance of the corresponding predictor $\tilde{Y}_n^{(j)}$: the expectation of the minimal (among h=1,...,k) squared prediction error is less than the minimal expectation of the squared prediction errors; the latter will, in turn, usually be less than the expectation of the squared prediction error of $\tilde{Y}_n^{(j)}$, as the probability of picking the best one is less than 1. Predicting the squared prediction error of $\tilde{Y}_n^{(j)}$ should take account of this. An obvious approach to do so is by cross-validatory assessment. For j=2 and 3, this will amount to "cross-validatory assessment of cross-validatory choice", cf. Stone (1974), p. 115. In principle, the proposals of Section 3 can be applied, $\tilde{Y}_n^{(j)}$ playing the role of $\hat{Y}_n(\hat{\theta}_{n-1})$. However, the problem of taking acount of the dimensionality of the predictor family, denoted p in Section 3, is now aggravated. The way of dealing with p in definitions (5) and (6) of the "conventional" and the first cross-validatory predictor of E_n^2 is based on the assumption that EE_n^2 is roughly proportional to (4). In the present situation, such an assumption may be shakier than it was already in Section 3: the dimensionality p is now ill-defined, as the dimensionalities of the k predictor families can be different, and the additional discrete parameter $h \in \{1,...,k\}$ has been introduced. The following two ad hoc approaches could be considered:

- define the dimensionality as $p = 1 + \max_h p_h$, where p_h is the dimensionality of the h'th predictor family (excluding blatantly unsuitable predictor families, if any, from this maximization);
- assume that EE_n^2 is proportional to (n-p+1)/(n-p) for some unknown p, and estimate p from the past realised prediction errors E_i^2 (i ≤ n-1);

in both approaches, definition (4) can be used for a "conventional" and definitions (5) and (6) for cross-validatory assessments of $\tilde{Y}_n^{(j)}$. The "conventional" predictor (4) is just the minimized $V_{nj}^{(h)}$ multiplied by a factor greater than 1.

Predicting the squared prediction error of $\tilde{Y}_n^{(j)}$ along these lines was not carried out in the reported simulation experiment. In any case, it is clear that the dependence on i of the distribution of E_i results in a lack of "balance" which poses even greater problems for cross-validatory assessment of cross-validatory choice than it does already for straight cross-validatory assessment.

5. DISCUSSION

In the framework of the predictor error method for single sime series, a comparison was made between a "conventional" and two cross-validatory procedures for prediction of squared prediction errors, and for choosing among several predictor families. The cross-validatory approaches considered were based on realized prediction errors in the past. It was found in a simulation study that the conventional procedure performed mostly somewhat better than the cross-validatory procedure for the purpose of prediction of squared prediction errors, while all three methods performed about equally well for the purpose of predictor family choice.

As in many simulation studies, it is unclear to which extent these results can be extrapolated to predictor families and time series distributions not considered in this simulation. All predictor families considered were of low dimensionality. The simulation results suggest that, provided dimensionality differences between predictor families are accounted for as in definitions (5) - (7), and whether or not the predictor family is explicitly or implicitly based on a correct model assumption for the time series, cross-validatory procedures are not better for predictor evaluation than conventional procedures. In particular these cross-validatory procedures did not prove to guard better against chance capitalization. This disappointing result concerning cross-validation is in line with some other results on risk functions of certain cross-validatory procedures; see e.g., Stone (1977).

These tentative conclusions do not detract from the possible usefulness of cross-validation for diagnostic purposes in the context of time series prediction. It seems very sensible to use the time series of past realized prediction errors E_1, defined as in (3), and past predictions V_{11} of squared prediction errors, defined in (5), to diagnose the performance of the predictor family for the time series at hand.

REFERENCES

Brent, R.P. (1974), A Gaussian pseudo random number generator. Comm.Ass. Comp. Mach. 17, 704-706.

Brown, R.L., Durbin, J., and Evans, J.M. (1975), Techniques for testing the constancy of regression relationships over time, Journal of the Royal Statistical Society, B, 37, 149-163.

Kendall, M., and Stuart, A. (1977), The advanced theory of statistics, Vol. 1, 4th Ed., Griffin, London.

Kendall, M., Stuart, A., and Ord, J.K. (1983), The advanced theory of statistics, Vol.3, 4th Ed., Griffin, London.

Ljung, L. (1978), Convergence analysis of parametric identification methods, IEEE Trans. Aut. Contr., AC-23, 770-783.

Ljung, L., and Söderström, T. (1983), Theory and practice of recursive identification, MIT Press, Cambridge, Massachusetts.

Shibata, R. (1985), Various model selection techniques in time series analysis, pp. 179-187 in Hannan, E.J., Krishnaiah, P.R., and Rao, M.M., eds., Handbook of Statistics, Vol. 5, North-Holland, Amsterdam.

Stone, M. (1974), Cross-validatory choice and assessment of statistical predictions, Journal of the Royal Statistical Society, B, 36, 111-147.

Stone, M. (1977), Assymptotics for and against cross-validation, Biometrika, 64, 29-35.

MODIFICATION OF FACTOR ANALYSIS MODELS IN COVARIANCE STRUCTURE ANALYSIS
A MONTE CARLO STUDY

Thom Luijben, Anne Boomsma, Ivo W. Molenaar[*]
University of Groningen
Vakgroep Statistiek en Meettheorie
Groningen, The Netherlands

ABSTRACT

   In covariance structure analysis statistics like the modification index and the "t-values" can be used for modifying a model. A model with an acceptable fit might be simplified, whereas a rejected model might be expanded or otherwise altered. This paper describes the results of Monte Carlo research in which the usefulness of a number of statistics is examined when a wrongly specified factor analysis model is modified in one parameter. For the samples where the correct model was obtained after a first modification step, the empirical distribution of the parameter estimates is examined. The conclusion is that low power and sampling variability in the decision criteria used for model modification, can be a serious threat to the reliability and validity of the decision which model to retain.

1. INTRODUCTION

   In causal modelling an investigator searches for a model which best fits the sample data. The problem of finding such a model can be solved in at least two ways. First, specify a set of nested models. By the use of a

   * This research was supported by the Foundation of Social-Cultural sciences which is subsidized by the Netherlands Organization for the Advancement of Pure Research (Z. W. O.) under project number 500-278-003. The authors wish to thank the editor for his helpful comments during the research and on earlier versions of this paper.

decision criterion these models can be tested and compared. From all these models the one which most adequately fits the sample data can be found. Secondly, a researcher can start with some baseline (starting, or initial) model. Often this baseline model does not fit the data well enough, or when it does, it might be too complex. Therefore the investigator decides to modify the model. This can be a parameter relaxation in order to get a better fit, or a model simplification to diminish its complexity. Other strategies involve specifying and comparing non-nested models. In practice, mixed strategies are frequently adopted.

This paper investigates the statistical properties of model modification but we want to stress in advance that whenever in practical research a model modification is proposed on statistical grounds, a check has to be made whether this modification is theoretically possible and, preferably, defensible. This study will mainly concentrate its attention at the comparing of nested models found by following the second procedure. Restriction to nested models diminishes the complexity of the problem. The reason for following the second procedure is that specifying and testing all possible nested models would be a very laborious job. Furthermore, on the basis of theoretical considerations a researcher has often evidence that a substantive number of models are impossible. For example, when one variable can only occur after the other in time, the direction of the direct cause is only possible in one way. Even stronger, for theoretical reasons it often occurs that the researcher has a testable baseline model (Saris, Zegwaart & De Pijper, 1979, p. 157; Cudeck & Browne, 1983, p. 151). Bentler & Bonett (1980, p. 596) and James, Mulaik & Brett (1982, p. 148) suggest to start with a very simple baseline model in contrast with Anderson (1962), Kennedy & Bancroft (1971) and Malinvaud (1980, chapt.7) who defend the thesis that it makes more sense to start with a general model.

The outline of the paper is as follows. In Section 2 we will briefly discuss the most convenient decision criteria which are used to test and/or to compare competing models in the LISREL framework (Jöreskog & Sörbom, 1984). Since 1973 LISREL programs are available which can be used to estimate parameters in a specified model and to examine the fit of that model. The program output gives statistics which can be used to decide whether and how a model has to be modified. In Section 3 the statistics under study, which are available in LISREL V and later versions, are described. In

Section 4 attention will be paid to the practical usefulness of these
statistics for modification of a certain factor analysis model by a Monte
Carlo study. We specify a true model and produce 300 sample covariance
matrices from it. Initially on each sample a wrongly defined model (one
parameter omitted) is fitted. For the samples for which the wrong model is
rejected, it is investigated which statistics can be used to detect the
correct model. Further it is investigated whether for the samples which
accept the wrong model the available statistics even suggest a simplification of the model. Next, following a suggestion of the editor, in Section 5
the empirical distribution of the parameter estimates will be examined for
the subclass of samples for which the true model has been found after one
modification step. Then the question is raised whether the value of the
specific parameter, defined in the true model but omitted in the initially
fitted model, will influence the capacity of the available statistics to
detect the true model. further, the influence of the value of this parameter
on the empirical distribution of the parameter estimates for the subclass of
samples for which the true model has been found after one modification step
is examined. These results are reported in section 6. Recently a comparable
study for a few structural models has been reported by MacCallum (1986). In
the discussion, which closes the paper, the results of this study are
compared with those of MacCallum.

2. STATISTICAL CRITERIA IN THE MODIFICATION PROCESS

In this section some statistical criteria, used to test and to compare
competing models, are briefly discussed. For a more extensive discussion we
refer to Luijben (1986a).

The likelihood ratio χ^2 test (Jöreskog & Sörbom, 1984, p. I.38) is
probably the most commonly used model testing procedure in covariance structure analysis. When the sample size is large enough, the test statistic has
approximately a noncentral chi-square distribution. The noncentrality
parameter (ncp) and therefore the test statistic, dependend on the sample
size except for the case where the true model has been detected, then the
ncp equals zero (Steiger, Shapiro & Browne, 1985, p.255). As acknowledged by
many researchers in the field, the fitted model is often, and probably
always, incorrect. Therefore, for large samples and a fixed significance

level virtually any model will be rejected, and in small samples various competing models are acceptable (Saris, Den Ronden & Satorra, 1987). To overcome this sample size dependency, fit indices were developed for which it was hoped the distributional dependency of the sample size was diminished. These measures for the overall fit are the goodness-of-fit index (GFI) and the root mean square residual (RMR). The GFI can be adjusted for degrees of freedom resulting in the adjusted GFI (AGFI) (Jöreskog & Sörbom, 1984, p. I.40). Both the AGFI and the GFI are transformations of the χ^2-value so it is doubtful whether these fit indices should be preferred to the usual χ^2-test statistic. Some other decision criteria which can be used to compare models are the normed fit index (Bentler & Bonett, 1980, p. 599) and the Akaike and Schwartz information criteria (Cudeck & Browne, 1983, p. 154). Relative to a baseline model, the normed fit index is the decrease in lack of fit between two nested models (James, Mulaik & Brett, 1982, p. 154). The information criteria combine a measure of fit and a measure of model complexity; they can be used for non-nested models too.

Finally, we mention the possibility of using cross-validation. To circumvent the difficulty of capitalization on chance and to assess the stability of a model, cross-validation can be very useful. A disadvantage of this procedure is the required sample splitting. Moreover, when the sample size increases, the models which survive cross-validation have an increased complexity (Cudeck & Browne, 1983, p. 162).

3. STATISTICAL CRITERIA FOR MODEL MODIFICATION

This section gives a description of the available statistical information which can be helpful in modifying a model. In practice a researcher accepts a model when it has a sufficient fit, or rejects it when it has not. When the model is accepted, it might be simplified while maintaining a sufficient fit; when it is rejected, it might be expanded or otherwise altered to reach a sufficient fit. In principle, two procedures can be used both for simplifying and for expanding a model. These procedures will be described first, followed by a description of statistical criteria which can be used only for expanding or only for simplifying a model.

I. The residual matrix. The residual matrix consists of the sample covariance matrix minus the fitted one. Costner & Schoenberg (1973, p. 177) have shown that an intuitive use of the residual matrix can lead to a completely wrong adaptation of the model. They propose a better method of inspecting the residual matrix, but this method is laborious and to a limited extent only available for factor analysis models. In practical research, LISREL is often used for more complex models than those considered in their paper. Therefore, using the residual matrix for the choice of a constraint which can be relaxed, or for fixing a parameter, appears problematic and will not be examined here.

II. The sequential χ^2 test and the sequential χ^2 difference test. An investigator using a sequential χ^2-test will modify his baseline model until the simplest model with an acceptable χ^2 when tested against a saturated model has been found. This is done by adding or deleting a parameter one by one. The sequential χ^2 difference test uses the difference in χ^2 between two models adjacent in the nested sequence as the criterion to stop or to continue modifying a model. When the difference in χ^2 between two models is significant, the more complex one will be chosen. After this step it is tested again whether adding another parameter will lead to a significant decrease in χ^2. For both methods it is therefore necessary that in each step two models are estimated. This paper concentrates on the choice of fixing or freeing a parameter before estimating a new model. The sequential χ^2 test and the sequential χ^2 difference test will therefore not be further discussed.

Statistical criteria for expanding a model

Three types of statistical information can be used for deciding in which direction a model has to be expanded.
(i) The first order derivatives (FD).
A fitting function measures in a way the discrepancy between the sample covariance matrix S and the estimated covariance matrix Σ, which is a function of the parameter estimates. The first order derivative of such a fitting function for a fixed parameter indicates the change in fit when this parameter is freed. In this study the FD of the loglikelihood function is examined. A large value of the FD indicates a large change in fit. Therefore, in the process of model modification the parameter with the largest (absolute) value of the FD could be freed.

(ii) The modification index (MI).

The modification index is an estimate for the decrease in χ^2 when a single parameter is relaxed without actually solving the new fitting problem (Sörbom, 1987; Jöreskog & Sörbom, 1984, p. I.42). The parameter with the largest MI will be the one which can freed. The value of that MI is important too. A small MI for a parameter suggests that, if this parameter is freed, no substantial decrease in χ^2 will occur. Jöreskog & Sörbom (1984, p. III.19) suggest to free a parameter when the value of its MI is larger than 5.

In LISREL the MI is given for every fixed parameter. For any fixed parameter which will cause identification problems in the model if this parameter is set free, the MI equals zero except for some pathological cases. Equivalently, the first order derivative for this parameter equals zero (Sörbom, 1987).

(iii) The expected parameter change (EPC).

For a parameter which is fixed at a certain value, frequently equal to zero, the expected parameter change can be calculated. The EPC is an estimate of the value of the parameter when it is freed, obtained without using a new iterative optimization procedure (Saris, Satorra & Sörbom, 1987). Using the LISREL VI program, the EPC equals: $-\text{MI}/[(\text{N}-1)\cdot\text{FD}]$. The fixed parameter with the largest absolute estimate can be chosen to be freed. How large this estimate of the parameter value has to be so that freeing this parameter is useful, is not quite clear, due to the scale dependence of the different parameters. For example, a change in the scale of the observed variables or the latent factors might cause a change in the values of the different EPC's. A preliminary result will be given indicating the effect of such a change.

Statistical criteria for simplifying a model

In practical research, a sample for which the fitted model is not rejected, can prompt the investigator to modify the model by fixing a free parameter at a certain value (Jöreskog & Sörbom, 1984, p. III.12). This value is often chosen as zero, meaning that a specific relation between variables in the model does not exist. For the sake of simplicity, this paper only discusses simplification of a model by fixing a parameter at zero and leaves out the possibilities of fixing a parameter at another value or constraining two parameters to be equal.

The investigator has to decide which parameter will be fixed. In the LISREL output the so-called "t-values" are given to indicate which parameters do not differ significantly from zero. A "t-value" is defined as the ratio of the parameter estimate and the corresponding estimated standard error. In practice, a parameter with a "t-value" smaller than 1.96, using a standard normal two sided test with $\alpha=0.05$, could be fixed. One could think that an estimate of the increase in χ^2 when a parameter is fixed, would add useful information in deciding whether a model can be simplified. However, such an estimate is just the square of the "t-value", and the critical "t-value" of 1.96 corresponds to the critical χ^2 of 3.84. Thus, the estimate of the χ^2 increase is no new information, and therefore this paper only uses the "t-value" as given in LISREL.

4. THE MONTE CARLO DESIGN

A simple fully identified model M_1 was chosen for which the known population parameter vector θ defines the population covariance matrix $\Sigma(\theta)$. From this model using the SIMLIS program (Boomsma, 1983), NR=300 sample covariance matrices S were taken, each with sample size N=200. A sample size of 200 and larger was chosen, because in practical research sample sizes of this order are often used. Also, for a sample size of 200 exact distributions appear to be rather well approximated by asymptotic distributions (Boomsma, 1983). Each sample was fitted by a model M_2 defined by a parameter vector, differing from θ by fixing one parameter. This was done because the purpose of the research is to investigate whether a wrongly specified, over-simplified, model can be corrected using statistical means. From the LISREL VI output for our 300 samples, the statistics under investigation were processed by a FORTRAN 77 program.

The population model M_1. In describing model M_1, the LISREL VI notation (Jöreskog & Sörbom, 1984, p. I.12) is used. Model M_1 consists of two factors (ξ_1, ξ_2), each factor having three indicators, (x_1, x_2, x_3) and (x_4, x_5, x_6), respectively. It is assumed that the vector $(x_1, x_2, x_3, x_4, x_5, x_6)$ has a multi-variate normal distribution with zero mean and covariance matrix $\Sigma(\theta) = \Lambda\Phi\Lambda' + \Psi$. The factor loadings Λ (6x2) are chosen in such a way that half of the observed variables have a nonzero loading on the first factor and a zero loading on the second one; the reverse holds for the other half. More specifically: $\Lambda' = \begin{pmatrix} 0.4 & 0.4 & 0.6 & 0 & 0 & 0 \\ 0 & 0 & 0 & 0.4 & 0.4 & 0.6 \end{pmatrix}$. The correlation matrix Φ of (ξ_1, ξ_2) has off-diagonal element ϕ; the value of this correlation between the two factors was chosen at 0.3. The covariance matrix of the errors of measurement Ψ (6x6 symmetric) is chosen in its most simple form: all the off diagonal elements are fixed at zero, and the diagonal elements are one minus the square of the corresponding loading. For example, $\psi_{11} = 1 - \lambda_{11}^2 = 0.84$ and $\psi_{66} = 1 - \lambda_{62}^2 = 0.64$. This implies that $\Sigma(\theta)$ has one's on its diagonal.

Model M_2 is defined as follows: Wherever the vector θ has a nonzero value, the corresponding parameter for M_2 is free, except for the correlation ϕ which is zero. Those parameters of θ which are fixed are also fixed in M_2. So M_1 and M_2 differ only through ϕ. The results obtained by this Monte Carlo study raised the question whether an increase in the value of ϕ would cause different results compared with the results reported in the next section. We therefore decided to repeat the Monte Carlo study for $\phi = 0.4$ and $\phi = 0.5$. These results are reported in section 6.

Results

To get 300 samples, a total of 309 samples was needed because of convergence problems for 9 sample covariance matrices. From those 300 samples, 27% rejected M_2 in favour of the saturated model using the likelihood ratio test at an α-level of 0.05. An approximation of the power of the likelihood ratio test gives 24% (Satorra & Saris, 1985, p. 83-90). The difference between the expected and the actual value of the power is thus small.

For brevity we will call a sample for which model M_2 is rejected a "rejection sample", and a sample which accept M_2 is called a "misleading sample". Before the rejection and misleading samples are discussed in detail, some comments have to be made about improper solutions (negative estimates of variances). In model M_2 this occurs when one of the ψ_{ii} is negatively estimated. This happened in 10% of all samples. There was one sample with negative estimates for two variances. For the samples where $\hat{\psi}_{ii}$ is negative, it has been suggested to fix this variance at a small positive value in order to retain the possibility of interpreting the results (Lawley & Maxwell, 1971, p. 32). We decided to ignore this suggestion, because in the view of the investigators, for the rejection samples the model will be expanded which results in new estimates, possibly without improper solutions. For the misleading samples, all the negatively estimated variances had the smallest "t-value", with one exception all smaller than 1.96. So this parameter is already the first candidate for fixing.

Now, first it will be shown which parameter relaxation for the 27% rejection samples is suggested by the modification criteria to be optimal, and next which simplification was most frequently suggested in the 73% misleading samples. In section 5, results on the distribution of the parameter estimates are given for samples for which the modification of model M_2 resulted in model M_1, with an acceptable fit.

The rejection samples

We will investigate the usefulness of FD, MI and EPC, mentioned in Section 3, for freeing a parameter in the 27% rejection samples. The question is whether parameter ϕ or another one will be proposed for freeing. Of course, the statistic which detects ϕ most often will be preferred, because this parameter is wrongly fixed.

In Figure 1 on the horizontal axis the parameters are given which lead to an identified model, when these parameters are allowed to vary individually. On the vertical axis for each of these parameters the percentage of the 80 rejection samples is displayed for which this parameter has the largest absolute FD, MI and EPC.

Inspection of·Figure 1 yields the following conclusions. Freeing a parameter as a consequence of using FD in the modification process does not seem to be very helpful. Approximately a uniform distribution of the FD across the parameters was found. Especially parameter ϕ was not chosen as often as expected.

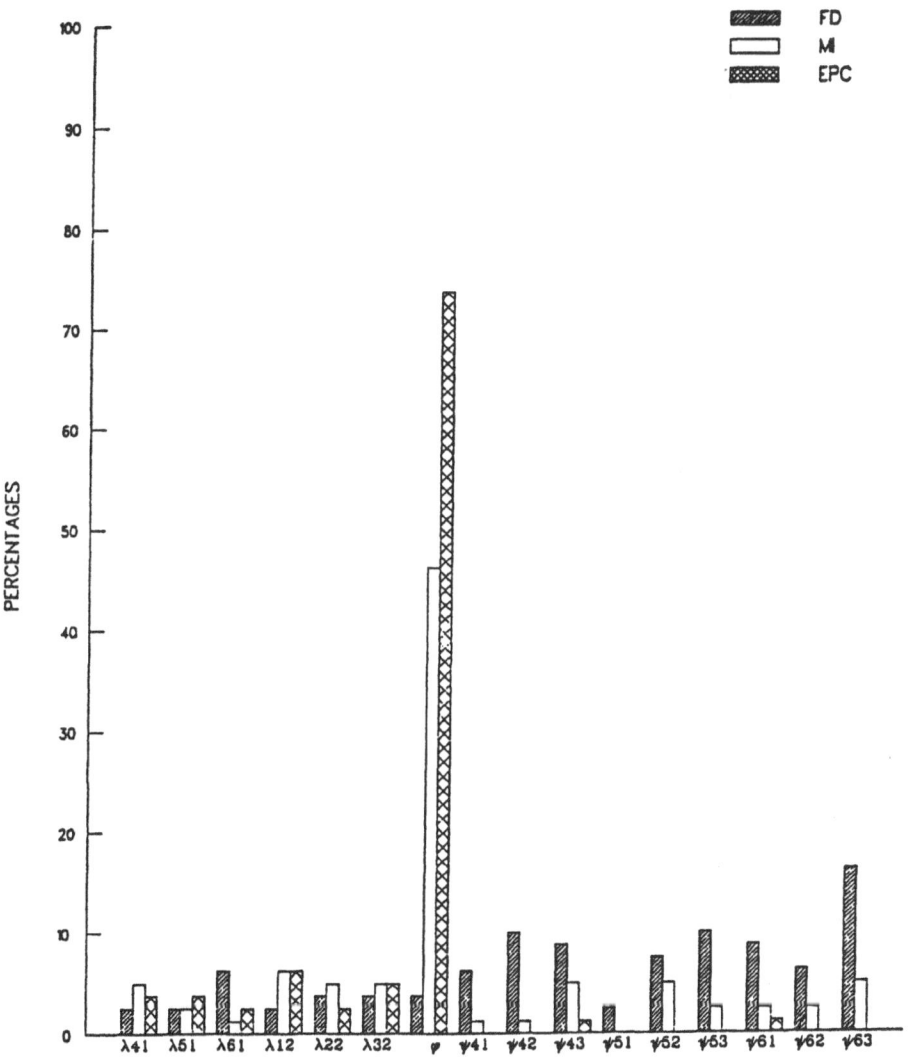

Figure 1. Comparing three procedures in a model modification process. The horizontal axis gives the parameters which when relaxed in M_2 lead to an identified model. The vertical axis gives the percentage of the 80 rejection samples for which a parameter has the largest absolute FD, MI and EPC, respectively.

The situation when the MI is inspected is completely different. Here parameter ϕ is in 43% of the rejection samples the one which should be freed. When the value of the MI is also considered, a decision has to be made which critical value the MI has to exceed before the corresponding parameter is expected to cause a "significant improvement" in fit when it is freed. Jöreskog & Sörbom (1984, p. I.42) note that under the null hypothesis the MI has asymptotically unconditionally a χ^2-distribution with one degree of freedom. With α=0.05, this means that the value of the MI has to be larger than 3.84. Inspecting the largest MI which also exceeds this critical value instead of inspecting only the largest MI, does not affect the decision in the rejection samples. The same holds when instead of 3.84 the value of 5 is used as proposed by Jöreskog & Sörbom.

The EPC seems to perform even better than the MI. With this criterion in 75% of the samples ϕ is chosen.

To answer the question whether these results could be expected in advance, we fitted model M_2 on the population covariance matrix $\Sigma(\theta)$. When model M_2 is fitted on this "theoretical" matrix, the "theoretical FD", the "theoretical MI" and the "theoretical EPC" for the free parameters can be obtained from the LISREL output.

To make a comparison of these "theoretical" values with the results obtained in this study, we use the (unconditional) results for all 300 samples.

It was expected that the parameter with the largest "theoretical FD" would also be the one which will appear most often as the parameter with the largest FD in the samples, since the FD's will be approximately normally distributed around the theoretical values. In Table 1a the "theoretical FD" values are given and in table 1b the percentage of the 300 samples for which the corresponding parameter has the largest absolute FD.

Table 1a. "theoretical FD values" for a parameter in the population.

param.	λ_{41}	λ_{51}	λ_{32}	λ_{12}	λ_{22}	λ_{61}	ϕ	ψ_{41}	ψ_{42}	ψ_{43}	ψ_{51}	ψ_{52}	ψ_{53}	ψ_{61}	ψ_{62}	ψ_{63}
minus FD	.04	.04	.07	.04	.04	.07	.07	.02	.02	.04	.02	.02	.04	.04	.04	.04

Table 1b. Percentage in the 300 samples for which a parameter has the largest absolute FD.

param.	λ_{41}	λ_{51}	λ_{32}	λ_{12}	λ_{22}	λ_{61}	ϕ	ψ_{41}	ψ_{42}	ψ_{43}	ψ_{51}	ψ_{52}	ψ_{53}	ψ_{61}	ψ_{62}	ψ_{63}
%	2	2	5	2	2	5	2	6	10	9	5	6	10	8	10	14

The results of the simulation study are quite opposite to what we expected: a high "theoretical value" does not lead to a high percentage.

The same comparison can be made for the "theoretical" values of the MI and the percentage for which a certain parameter has the largest MI in the 300 samples. The "theoretical values" for the MI based on the population covariance matrix $\Sigma(\theta)$ are presented in Table 2a and the percentage of the 300 samples for which the corresponding parameter has the largest MI, is given in table 2b.

Table 2a. "theoretical MI values" for a parameter in the population.

param.	λ_{41}	λ_{51}	λ_{32}	λ_{12}	λ_{22}	λ_{61}	ϕ	ψ_{41}	ψ_{42}	ψ_{43}	ψ_{51}	ψ_{52}	ψ_{53}	ψ_{61}	ψ_{62}	ψ_{63}
MI	.49	.49	1.82	.49	.49	1.92	4.22	.06	.06	.21	.06	.06	.21	.21	.21	.79

Table 2b. Percentage in the 300 samples for which a parameter has the largest MI.

param.	λ_{41}	λ_{51}	λ_{32}	λ_{12}	λ_{22}	λ_{61}	ϕ	ψ_{41}	ψ_{42}	ψ_{43}	ψ_{51}	ψ_{52}	ψ_{53}	ψ_{61}	ψ_{62}	ψ_{63}
%	4	3	6	5	5	9	35	4	5	6	2	4	4	3	3	4

The MI has a noncentral χ^2 distribution with one degree of freedom and a noncentrality parameter (ncp), which is estimated by the value of the MI when the population covariance matrix is fitted. Therefore, the expected value of the MI equals ncp+1, and its variance is 4ncp+2. Parameter ϕ is expected to appear frequently as the one with the largest MI.

Finally, we can make the same comparison for the "theoretical" values of the EPC and the percentage for which a certain parameter has the largest EPC in the 300 samples. The "theoretical values" for the EPC based on the population covariance matrix $\Sigma(\theta)$ are presented in Table 3a and the percentage for which a certain parameter has the largest EPC in the 300 samples is given in table 3b.

Table 3a. "theoretical EPC values" for a parameter in the population.

param.	λ_{41}	λ_{51}	λ_{32}	λ_{12}	λ_{22}	λ_{61}	ϕ	ψ_{41}	ψ_{42}	ψ_{43}	ψ_{51}	ψ_{52}	ψ_{53}	ψ_{61}	ψ_{62}	ψ_{63}
EPC	.07	.07	.13	.07	.07	.13	.30	.02	.02	.03	.02	.02	.03	.03	.03	.06

Table 3b. Percentage in the 300 samples for which a parameter has the largest EPC.

param.	λ_{41}	λ_{51}	λ_{32}	λ_{12}	λ_{22}	λ_{61}	ϕ	ψ_{41}	ψ_{42}	ψ_{43}	ψ_{51}	ψ_{52}	ψ_{53}	ψ_{61}	ψ_{62}	ψ_{63}
%	5	5	6	5	4	4	62	0	1	1	1	1	1	1	1	0

We could have expected beforehand that the parameter ϕ will most frequently have the largest EPC value.

We think that our model has no large differences in scale between the parameters, because all variables in the population have a variance of one. It is unclear whether the better performance of EPC, compared with MI, is due to a structural difference between the two statistics, to the chosen model or to chance.

As suggested in the introduction, a change in the scale of the observed or latent variables might influence the values of the EPC. To investigate this we fitted a model equivalent to model M_2 on the same 300 samples. In this model the parameters λ_{31} and λ_{62} were fixed at 1.0, and ϕ_{11} and ϕ_{22} were estimated freely. This implies that we change the scale of the latent factors. Both ξ_1 and ξ_2 will have now a variance of .36. The "theoretical MI values" for the fixed parameters did not change in contrast with the "theoretical EPC values". The "theoretical EPC value" was now .22 for λ_{61} but only .11 for ϕ. In the 300 samples the results were exactly the same with respect to the MI but changed dramatically with respect to the EPC: in none of the rejection samples ϕ had the largest absolute EPC. This leads to the conclusion that at least an automatic use of the EPC for model modification has to be dissuaded.

Conclusion. Both MI and EPC seem to be useful for finding the correct parameter which can be freed. EPC seems to work the best in this design

where the variances of the observed and latent variables are equal. The
value of the FD procedure seems very doubtful, because across samples almost
any parameter may be chosen to be freed, and especially ϕ is chosen too
infrequently. From a theoretical point of view the FD can better be
investigated in the way suggested by Byron (1972). Then it is not the
absolute value of the first order derivative which is inspected, but the so-
called "t-value", which is the same as the square root of MI. As we have
seen, there are statistical criteria for using MI. Thus consideration of MI
is preferred to a heuristic investigation of the value of the first order
derivative as presented here.

The misleading samples

For each sample of the 73% misleading ones, an investigator might
decide to simplify the model by fixing a parameter. As explained in Section
3, inspecting the "t-values" of the free parameters seems to be useful. In
45% of the misleading samples (33% of the total of 300) there was no
parameter which had a "t-value" smaller than 1.96. In practice, in such
cases an investigator will often be satisfied, because the fit is acceptable
and there is no parameter which seems reasonable to fix. So for no less than
33% of all samples the investigator retains the wrong model.

For 55% of the misleading samples, there was at least one free
parameter which had a "t-value" smaller than 1.96. The percentage for each
parameter is given in Figure 2.

The parameters ψ_{33} and ψ_{66} are the ones which frequently had the
smallest "t-value"; 41% and 32%, respectively. The question is whether such
results could be expected. The "theoretical t-values" based on the
population covariance matrix $\Sigma(\theta)$ given in table 4, are an indication for
the frequency that a certain parameter will have a small "t-value" in a
certain sample out of the 300 samples.

Table 4. "theoretical t-values" for the parameters in the population.

param.	ψ_{11}	ψ_{22}	ψ_{44}	ψ_{55}	λ_{31}	λ_{62}	ψ_{33}	ψ_{66}	λ_{11}	λ_{21}	λ_{42}	λ_{65}
t-val.	7.40	7.40	7.40	7.40	3.80	3.80	3.46	3.46	3.40	3.40	3.40	3.40

The mean "t-values" in the 300 samples and in the 220 misleading ones are roughly equal to these theoretical values. The fact that the parameters ψ_{33} and ψ_{66} have the smallest "t-value" so frequently within the subset of the misleading samples can be explained by the interpretation of the variances of the "t-values" for each parameter. The parameters ψ_{33} and ψ_{66} have a relative large variance (about 5) in comparison with the other parameters (variances of about 1.5 for the parameters which had a "t-value" about 3.4 and about 3 for the parameters which have the large "t-values" of about 7).

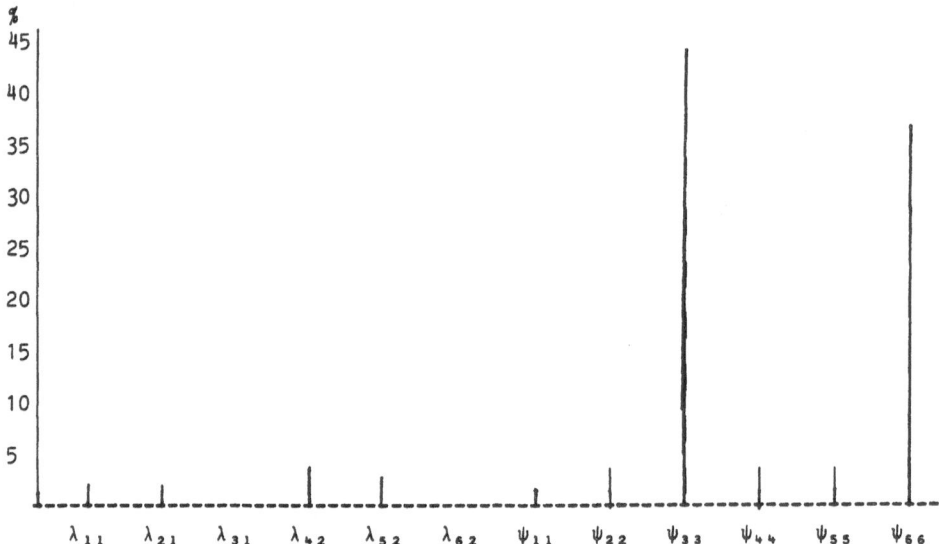

Figure 2. Simplifying M_2 by fixing a free parameter at zero. On the horizontal axis the free parameters of M_2. On the vertical axis the percentage for which that parameter has the smallest "t-value" of the misleading samples which have at least one "t-value" smaller than 1.96.

Conclusion. For almost half of the misleading samples an investigator will be satisfied with model M_2. For the other misleading samples an investigator might even try to further simplify the model. The parameters ψ_{33} and ψ_{66} will then most frequently be fixed. In accordance with the recommendation given in the introduction, an investigator might doubt whether such a simplification is theoretical plausible. for example, fixing ψ_{33} at zero means that the factor ξ_1 is explained perfectly by variable x_3, which can be an unrealistic assumption.

5. THE EMPIRICAL DISTRIBUTION OF PARAMETER ESTIMATES AFTER A SUCCESSFUL ONE STEP MODIFICATION PROCEDURE

For all the rejection samples where parameter ϕ is chosen to be freed, the true model M_1 is found. In this section we examine the empirical distribution of the parameter estimates under M_1 based on these samples. We want to find out whether the modification process influences the distribution of the parameter estimates. The reason is that in practice the estimates are evaluated ignoring the modification process, i.e. one use unconditional distributions to asses their accuracy. But the sample at hand belongs to a subset of the sample space, characterized by the property that they all lead, through the adopted modification process, to the true model. A priori it is not clear that the conditional and unconditional distributions are identical.

When the observed variables are multivariate normally distributed and the sample sizeislarge enough, the standardized parameter estimates $q(\theta)=(\hat{\theta}-\theta)/se(\hat{\theta})$ have approximately a standard normal distribution.

It could be questioned whether the asymptotic theory is valid for the sample size of 200 used in this study. To check this, the empirical distribution of the standardized parameter estimates is inspected when model M_1 is fitted on all 300 sample covariance matrices. In Figures 3a and 3b the histograms and corresponding QQ-plots are given for the standardized parameter estimates $q(\lambda_{11})$ and $q(\phi)$, which are presented here as examples for the free parameters. We choose $q(\phi)$ for the purpose described below and $q(\lambda_{11})$ as a representation of the other parameters.

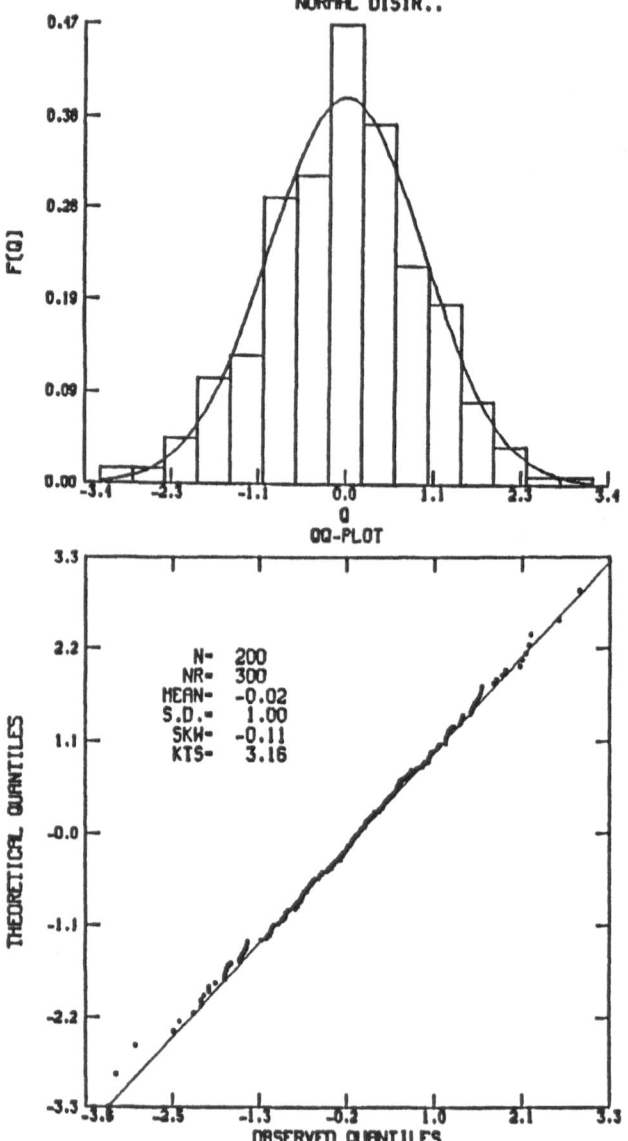

Figure 3a. Histogram and QQ-plot of the empirical sampling distribution of the standardized parameter estimate $q(\lambda_{11})$ for all 300 samples.

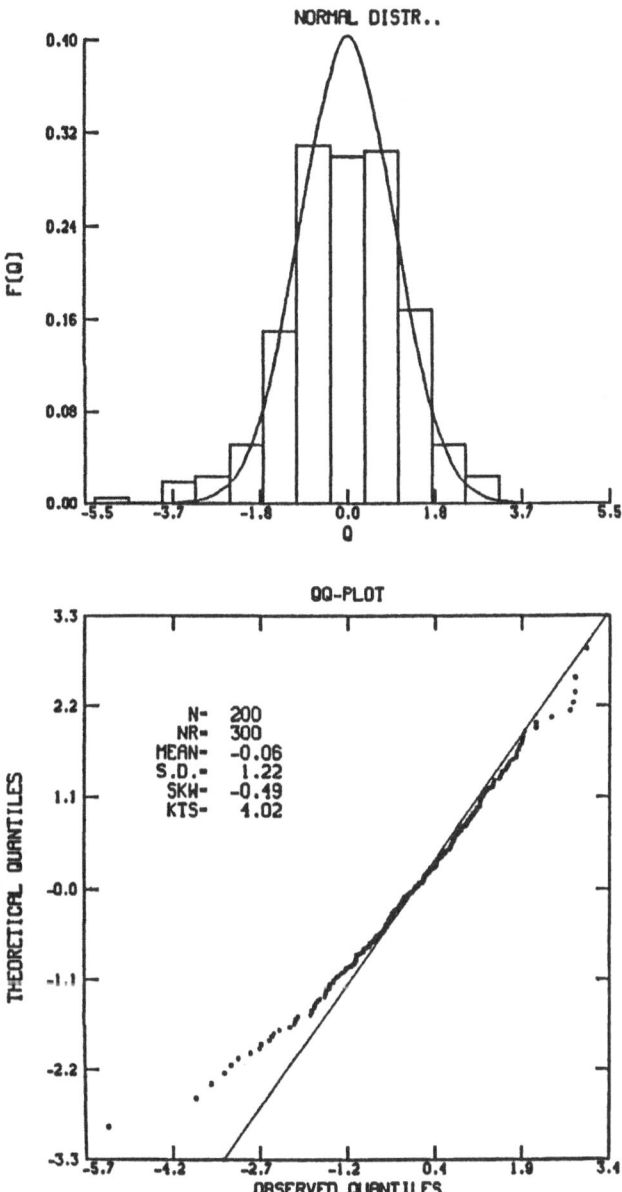

Figure 3b. Histogram and QQ-plot of the empirical sampling distribution of the standardized parameter estimate q(ϕ) for all 300 samples.

Inspecting these figures warrants the conclusion that no substantial deviations from the normal theory with respect to the mean occur. More formally, univariate hypotheses tests (one-sample Student t-test) were used to examine whether the mean of the standardized parameter estimates deviates from zero. Table 5a gives the resulting test statistics.

Table 5a. Student t-test for the mean of standardized parameter estimates $q(\theta)$ in model M_1 based on all 300 samples.

param.	λ_{11}	λ_{21}	λ_{31}	λ_{42}	λ_{52}	λ_{62}	ϕ	ψ_{11}	ψ_{22}	ψ_{33}	ψ_{44}	ψ_{55}	ψ_{66}
t-test	-.40	1.16	-3.39	.25	.02	4.01	-.85	1.20	1.45	2.88	1.91	-2.03	3.83

For some parameters a deviation from the expected value occurs. These deviations may be caused by the moderate sample size of 200, but also by outliers and the fact that in a set of 13 statistics like these, chances are good to find at least one large value. For an exhausting discussion about deviations from normal theory for this kind of models and similar sample sizes, we refer to Boomsma (1983).

The main purpose of this section is to compare the empirical sampling distributions based on the 300 samples with those for the subset of samples for which the true model has been found after one modification step. It is possible to choose this subset in three different ways: those rejection samples for which the FD, MI, and EPC of parameter ϕ, respectively, is the largest. As discussed in the previous section, the FD approach does not seem to be a reasonable choice. The choice between the two other approaches is more difficult. We decided to examine the 59 rejection samples for which ϕ has the largest EPC, because the EPC performs best, and thus there is a larger subset than according to the MI approach (37 samples).

Fitting the true model M_1 on these 59 sample covariance matrices resulted in one sample with a convergence problem, five rejection samples in favour of the saturated model, and 53 misleading ones, at a significance level of 5%. When a one modification step procedure is followed, an investigator will be satisfied with model M_1 when the sample has an acceptable fit. Therefore, the subset of 53 samples which accept M_1 will be examined. For these 53 samples histograms and QQ-plots of the empirical distribution of the standardized parameters $q(\lambda_{11})$ and $q(\phi)$ are displayed in Figures 4a and 4b, respectively.

Inspecting the figure, at first sight for $q(\lambda_{11})$, there are no substantial differences from the (hoped for) standard normal behavior. Probably due to the small number of samples the empirical sampling distribution of $q(\lambda_{11})$ is somewhat more scattered than the corresponding one based on the total of 300 samples (compare Figure 3a with 4a). The same result holds for the other parameters (not displayed here), except for ϕ. In Figure 3b it can be seen that the empirical sampling distribution of $q(\phi)$, based on 300 samples, has approximately a normal distribution. For the subset of 53 samples a normal distribution also fits approximately, but with a mean equal to 1.23 and a standard deviation equal to .83 (see figure 4b). The mean and standard deviation of the 53 values $\hat{\phi}$ equal .473 and .122 respectively.

It is illustrative to present the results of one-sample Student t-tests on the means of standardized parameters $q(\theta)$. The test statistics are given in Table 5b.

Table 5b. Student t-test for the mean of standardized parameter estimates $q(\theta)$ in model M_1 based on the 53 successful samples.

param.	λ_{11}	λ_{21}	λ_{31}	λ_{42}	λ_{52}	λ_{62}	ϕ	ψ_{11}	ψ_{22}	ψ_{33}	ψ_{44}	ψ_{55}	ψ_{66}
t-test	2.44	1.63	-1.38	-.71	1.30	-.89	10.8	4.03	-.14	2.0	-1.76	-1.34	-.04

The findings seem to justify the conclusion that the empirical sampling distribution of standardized parameter estimate $q(\phi)$ is strongly affected by the outcome of the modification process. When comparing the results from Table 5a with those from Table 5b except for parameter ϕ no dramatic differences are present. It is therefore concluded that mainly $q(\phi)$'s distribution is affected.

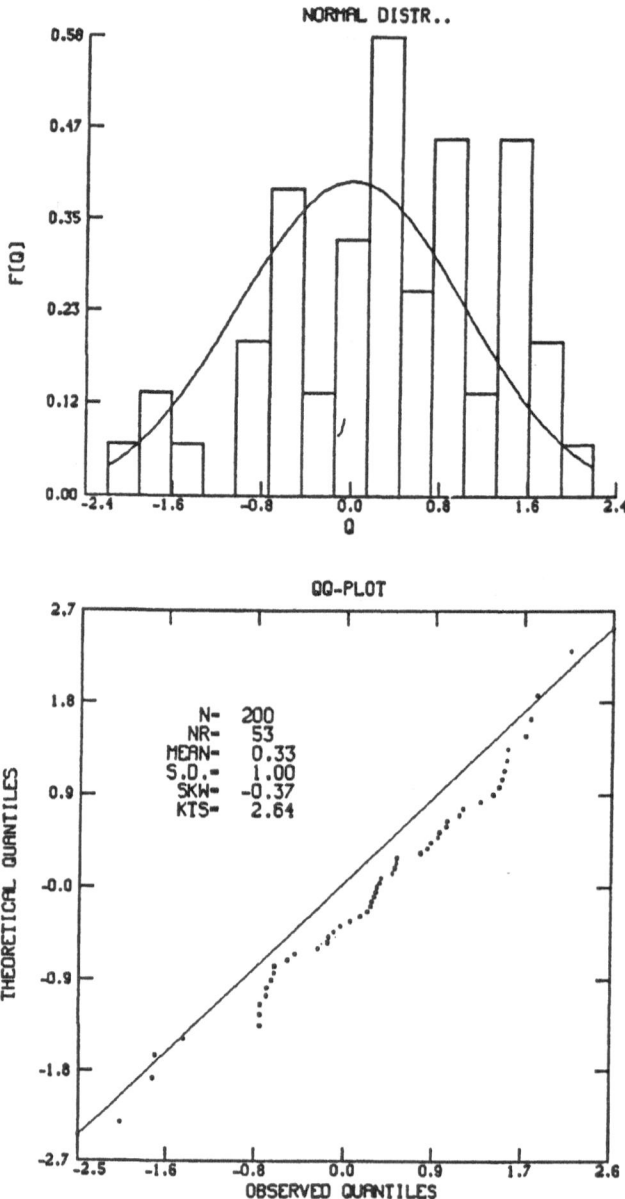

Figure 4a. Histogram and QQ-plot of the empirical sampling distribution of the standardized parameter estimate $q(\lambda_{11})$ for the 53 successful samples.

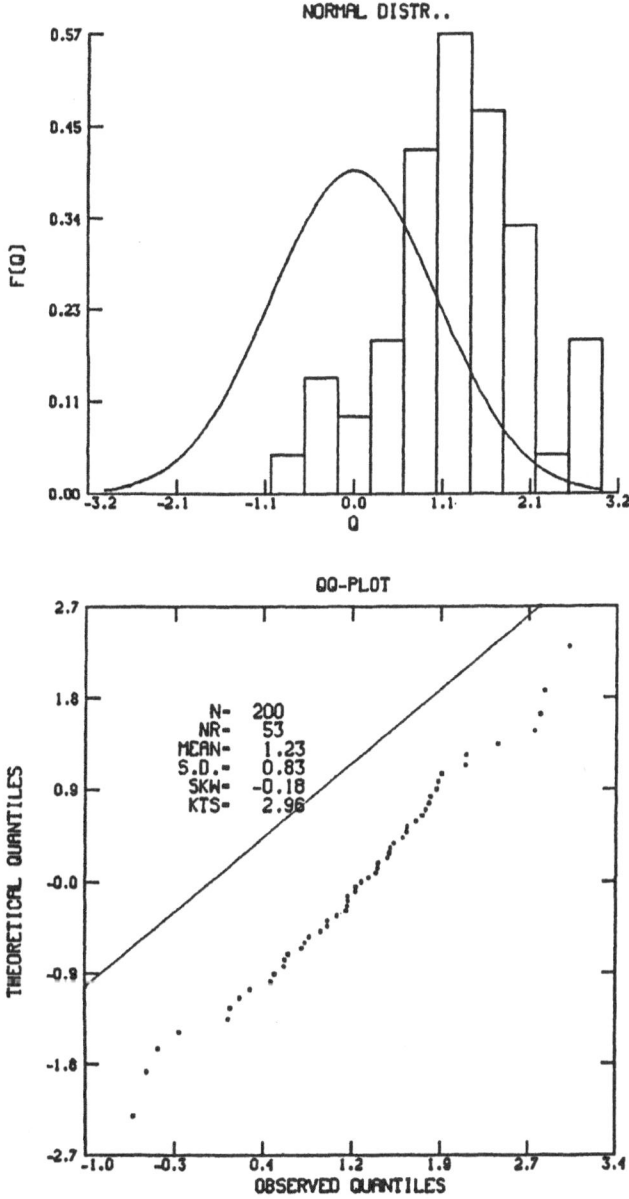

Figure 4b. Histogram and QQ-plot of the empirical sampling distribution of the standardized parameter estimate q(ϕ) for the 53 successful samples.

The shift in the distribution of $q(\phi)$ was not really suprising. When M_2 was fitted to the population covariance matrix $\Sigma(\theta)$, except for ϕ the parameter estimates turn out to be exactly equal to those when model M_1 was fitted. In other words, freeing ϕ (going from model M_2 to M_1) does not affect the parameter estimates based on the population matrix. This phenomenon is expected approximately to persist in each sample. Keeping this in mind may help to understand the following explanation of the noticed shift in $\hat{\phi}$. For the 59 rejection samples the χ^2-value of the likelihood ratio test is relatively large when M_2 is fitted; the minimum value is 16.9. Because of the direct relationship between the fitting function and the residuals, this implies that when model M_2 is fitted, the residuals are relatively large. This is mainly due to the absence of ϕ. As a result of the dependence of the parameter estimate ϕ from the residuals, in general, the estimates of $q(\phi)$ in the 53 samples will be larger than the comparable ones in the 300 samples.

When the power of the test increases, we expect a decrease of the impact of the modification process on the distribution of the parameter estimates. For example, a larger sample size, a larger population value of ϕ or larger loadings will increase the power of the test, in general. This will lead to a larger number of rejection samples when Model M_2 is fitted, and therefore an increase of the number of samples which will find the correct model after one modification step. The larger this subset of samples, the smaller the shift in $\hat{\phi}$ will be. In section 6, we want to investigate one of these hypotheses: will the shift in the distribution for $q(\phi)$ decrease when the value of ϕ in the population model increases ?

Conclusion. For this model, the estimates of parameter $q(\phi)$ are substantially affected by the modification process whereas the other parameter estimates are not.

6. THE BEHAVIOUR OF THE MODIFICATION STATISTICS AND THE EMPIRICAL DISTRIBUTION OF THE STANDARDIZED PARAMETER ESTIMATES AFTER A ONE STEP MODIFICATION PROCESS WHEN THE VALUE OF ϕ IN THE POPULATION INCREASES.

In this section two aspects of the modification process are investigated: first: the behaviour of the modification statistics when the value of ϕ increases in the population. For larger values of ϕ, the power increases. We want to know whether the MI and the EPC will then more frequently point at ϕ as the missing parameter. Investigations about the FD are omitted, because this statistic did not function well. Also the "t-values" are not inspected for reasons of brevity.

Second, the question will be answered whether an increase in the value of ϕ will lead to an decrease in the shift for the empirical distribution of the standardized estimates $q(\phi)$. This is expected for reasons given in the last part of section 5.

We decided to repeat The Monte Carlo design and produced therefore sample covariance matrices from a population matrix defined by the same values for the loadings and the uniqhesses, but with a value for ϕ of 0.4 and 0.5 respectively.

Results.

Results are reported for $\phi=0.4$, with the corresponding results for $\phi=0.5$ between brackets.

To get 300 samples, a total of 313 (304) samples was needed because of convergence problems. From the 300 samples, 52% (70%) rejected model M_2 in favour of the saturated model using the likelihood ratio test at an α-level of 0.05. An approximation of the power of the likelihood ratio test gives 48% (69%) (Satorra & Saris, 1985, p. 83-90). The difference between the expected and the actual value of the power is again small.

To answer the question whether the MI and the EPC function better in the detection of ϕ, figure 5a presents the performance of MI for the rejection samples based on both $\phi=0.3$ and $\phi=0.5$. Figure 5b similarly presents the performance of EPC.

94

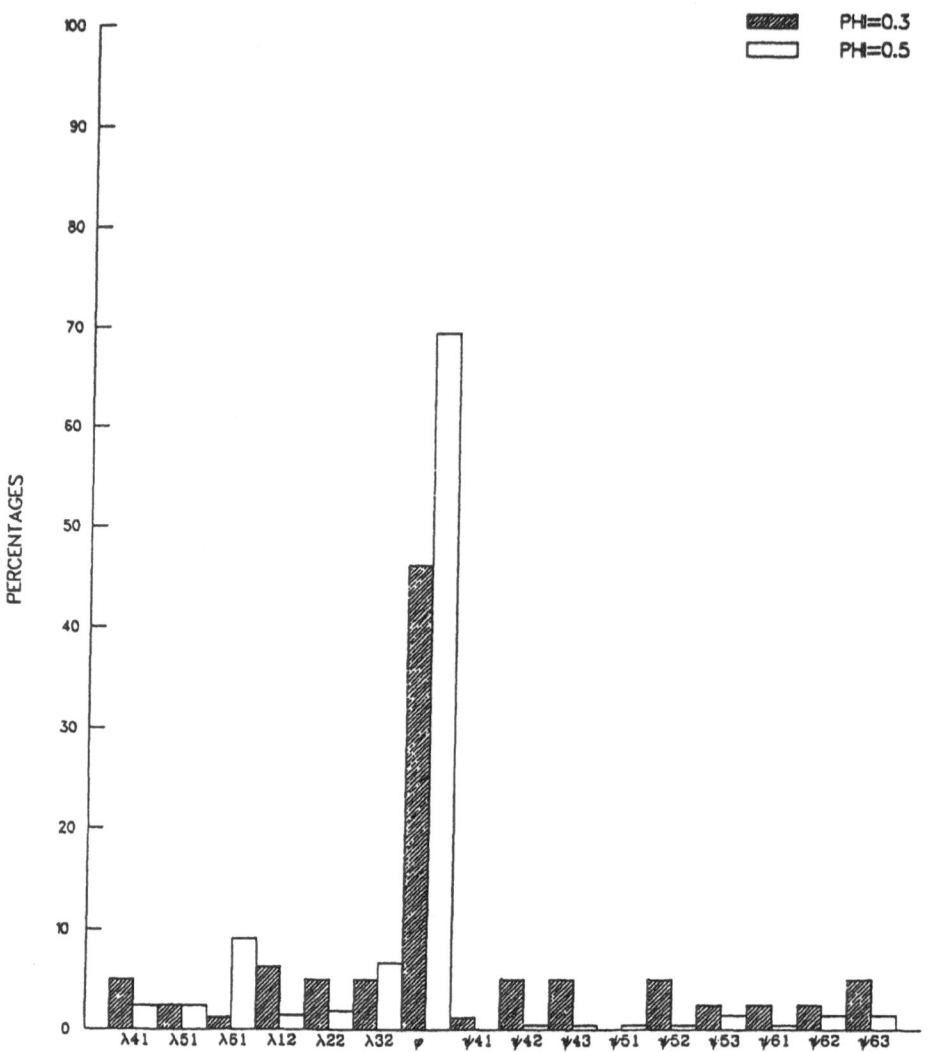

Figure 5a. Comparing the MI's for two population values of parameter φ. The horizontal axis gives the parameters which, when relaxed in M₂, lead to an identified model. The vertical axis gives the percentage of the 27% and 70% rejection samples for which a parameter has the largest MI.

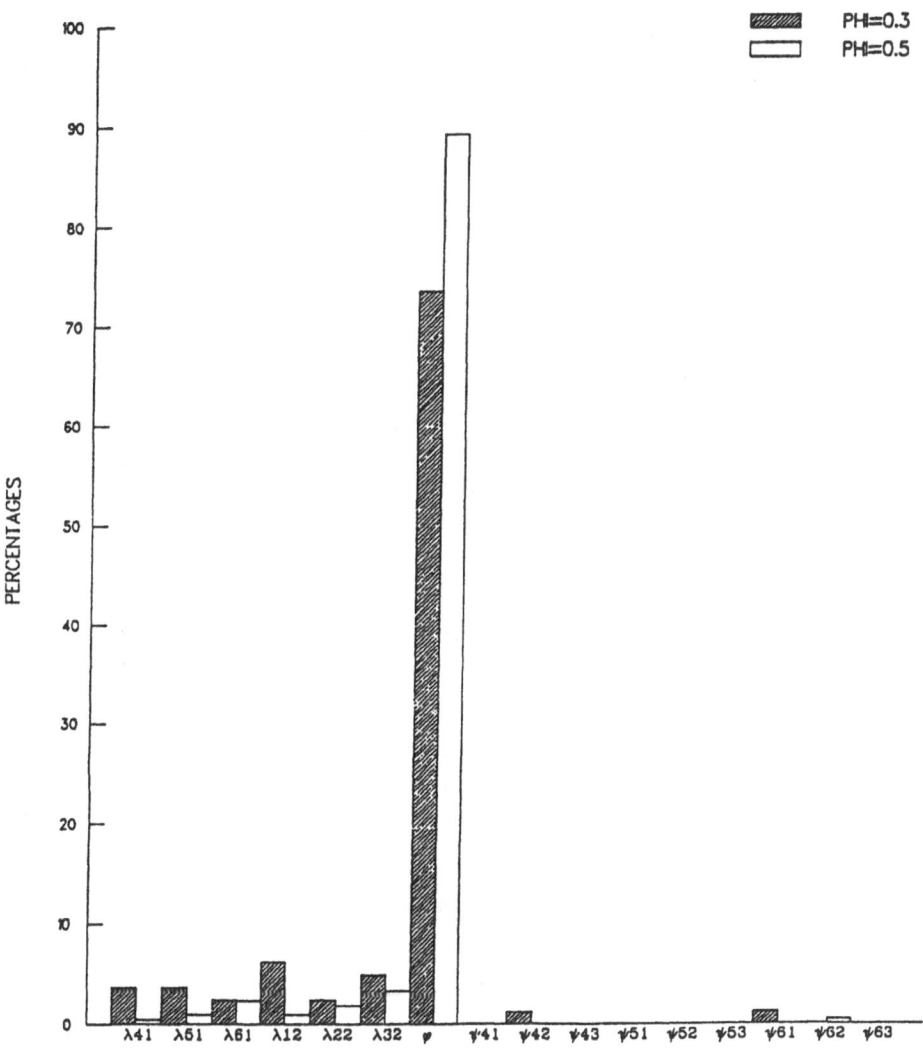

Figure 5b. Comparing the EPC's for two population values of parameter ϕ. The horizontal axis gives the parameters which, when relaxed in M_2, lead to an identified model. The vertical axis gives the percentage of the 27% and 70% rejection samples for which a parameter has the largest EPC.

Both figures lead to the conclusion that the MI and EPC function relatively better when the population value of ϕ increases. For the design with $\phi=0.4$, the functioning of MI and EPC can be characterized as "intermediate between $\phi=0.3$ and $\phi=0.5$". For example, MI and EPC pointed at ϕ in 59% (MI) and 81% (EPC) of the rejection samples.

Finally, it is investigated whether the noticed shift in the empirical distribution of the standardized parameter estimates of $q(\phi)$ following the described test-retest procedure, changes as a function of the population value ϕ. There were 157 (210) samples for which model M_2 was rejected. The EPC indicated that in 129 (188) of these samples, ϕ had to be set free. Fitting M_1 on these samples resulted in 121 (180) misleading ones for which the mean of these estimates of ϕ was .556 (.567). In figure 6a (6b) the QQ-plots for the empirical distribution of the standardized parameter estimates of $q(\phi)$ for these samples is given.

Comparing the plots of figure 4b, 6a and 6b show that an increase of the population value of ϕ causes a smaller distance between the plotted points and the theoretical line. This confirms the expectation that an increase of the population value of ϕ causes a decrease in the shift.

To make this comparison more strict, the one-sample Student t-test values on the mean of the $q(\phi)$'s were 10.8, 15.0 and 7.6 for ϕ equal to 0.3, 0.4 and 0.5 respectively. Still, these t-values are all significant. It was expected that the t-values would also decrease with the increase of ϕ; the t-value of 15 is therefore higher than expected.

7. DISCUSSION

The results of this Monte Carlo study are not very encouraging. Although started with a simple model which differs in only one parameter from the true one, a one step modification procedure results for only 53 out of 300 samples in the detection of the correct model with an acceptable fit. In addition, the empirical distribution of the standardized parameter estimate for the specific parameter, omitted from the correct model, showed a shift

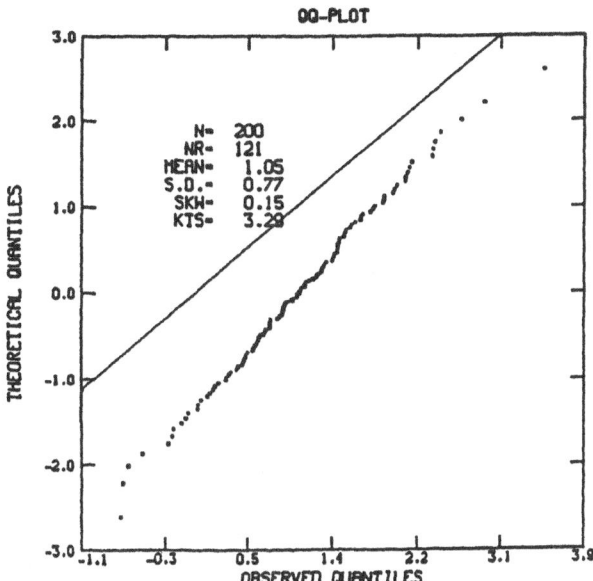

Figure 6a. QQ-plot of the empirical sampling distribution of the standardized parameter estimate q(φ) for the 121 successful samples for a population
value φ=0.4.

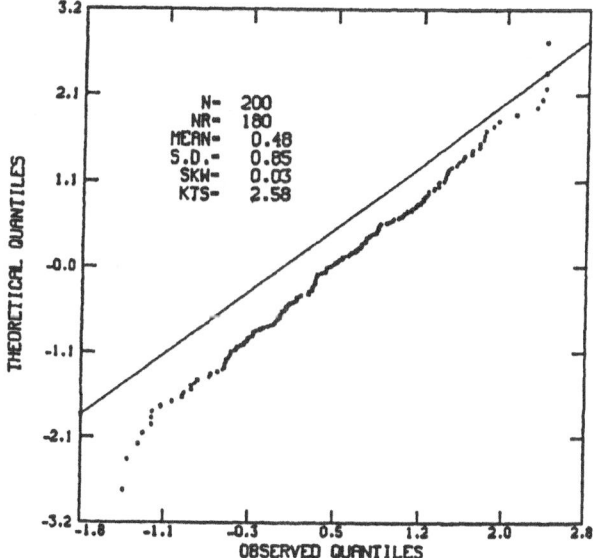

Figure 6b. QQ-plot of the empirical sampling distribution of the standardized parameter estimate q(φ) for the 180 successful samples for a population
value φ=0.5.

and an incorrect scale. This means that when an investigator finds the correct model, a bias in the estimate for the added parameter must be expected and confidence intervals will be to short and out of focus.

The main reason for this meager result is a low power of 27%. Furthermore, not for all rejection samples, the correct modification of model M_2 has been found, using the available statistics. When the modification process would be continued, it is expected that the result would even be worse. The samples which accept M_2 after the first step, will not further be modified or will even be further simplified. For the samples which found M_1 after one modification step, maybe a more simple model with an acceptable fit can be found. The additional relative number of samples for which the true model is found after more than one modification step, is expected to be small.

The results become better when the population value of the crucial parameter increases. This leads directly to a higher power, but also to a better functioning of the MI and the EPC. When the value of this parameter would further increase, it is expected that the detection of this parameter for both the MI and the EPC, will converge to 100%. Also the noticed shift will decrease and, of course, will vanish when the value of the parameter will be so high that the power is 100%. It is an interesting question whether these results will hold in other models. A useful comparison can be made with the study of MacCallum (1986). He studied a number of structural models, but with a small number of replications (NR=20) for each model. The small number of replications has the advantage that the modification process for each sample can be examined in more steps until the most acceptable model has been found. The advantage of our design is that more information can be gathered about the percentage of samples which are rejections, the percentage of correct modification decisions, and the empirical sampling distributions of the parameter estimates after a modification process. The main conclusion of MacCallum was that under unfavourable conditions, such as a baseline model which differs much from the true model, small sample size and the fact that investigators are satisfied too quickly with a model with an acceptable fit, models arising from modification processes must be viewed with caution, because they are likely to be invalid and unstable.

We have seen that an increase in the value of the crucial parameter leads to better results. A next step in this kind of research could be the investigation whether an increase in the sample size or higher factor loadings, which is expected to result in higher power (Satorra & Saris, 1985), will also show a relative better functioning of the MI and EPC in detecting the omitted parameter and a decrease in the shift. One could also investigate whether more indicators per factor would cause better results.

Finally a suggestion will be made for finding an estimate of the noticed shift. Suppose the sample at hand rejects M_2 and accepts M_1 (where $\phi=0.3$). When we fit M_1, we expect a bias in the estimate of ϕ. Suppose this estimate equals 0.43. Now use the bootstrap to produce 200 replications with a sample size of 200. Fit on these 200 replications M_2. For the subset of samples which reject this model, fit M_1. Calculate for these samples the mean of the estimates of ϕ. Suppose this mean equals 0.54. A hypothesis is now that 0.54-0.43=0.11 is a unbiased estimate of the bias. A researcher could therefore correct his estimate of ϕ from 0.43 in 0.43-0.11=0.32. Clearly, this procedure needs a lot of testing before it can safely be recommended; but it could be worthwhile since alternatives are not exactly abundant. However, the results obtained by Dijkstra & Veldkamp and Freedman et al. concerning the bootstrap and model uncertainty (this volume) are not very encouraging.

8.REFERENCES

Anderson, T. W., (1962). The choice of the degree of a polynomial regression as a multiple decision problem. Annals of Mathematical Statistics, 33, 255-265.

Bentler, P., & Bonett, D. (1980). Significance test and goodness of fit in the analysis of covariance structure. Psychological Bulletin, 88, 588-606.

Boomsma, A., (1983). On the robustness of LISREL (maximum likelihood estimation) against small sample size and nonnormality. Amsterdam: Sociometric Research Foundation.

Byron, R., (1972). Testing for misspecification in econometric systems using full information. International Econometric Review, 13, 745-756.

Costner, H., & Schoenberg, R. (1973). Diagnosing indicator ills in multiple indicator models. In A.S. Goldberger & O. Duncan (Eds.). Structural equation models in the social sciences. New York: Seminar Press, 168-200.

Cudeck, R., & Browne, M. (1983). Cross validation of covariance structures. Multivariate Behavioral Research, 18, 147-167.

James, L., Mulaik, A. & Brett, J. (1982). Causal analysis. Beverly Hills: Sage.

Jöreskog, K.G., & Sörbom, D., (1984). Lisrel VI: Analysis of linear structural relationships by the method of maximum likelihood: User's guide. Mooresville, Indiana: Scientific Software.

Kennedy, W. J. & Bancroft, T. A., (1971). Modelbuilding for prediction in regression based upon repeated significance tests. Annals of Mathematical Statistics, 42, 1273-1284.

Lawley, D.N., & Maxwell, A.E., (1971). Factor analysis as a statistical method. London: Butterworth.

Luijben, T.C., (1986a). Modification of models. Heymans Bulletin, HB-86-805-EX Rijksuniversiteit Groningen, Vakgroep Statistiek & Meettheorie.

Luijben, T.C., (1986b). Modification of factor analytical models in covariance structure analysis. Heymans Bulletin, HB-86-813-EX. Rijksuniversiteit Groningen, Vakgroep Statistiek & Meettheorie.

MacCallum, R., (1986). Specification searches in covariance structure modeling. Psychological Bulletin, 100, 107-120.

Malinvaud, E., (1980). Statistical methods of Econometrics, North-Holland: Amsterdam.

Saris, W., Zegwaart, P., & De Pijper, W., (1979). Detection of specification errors. In D.R. Heise (Ed.). Sociological methodology 1979. San Fransisco: Jossey-Bass, 151-171.

Saris, W., & Stronkhorst, H., (1984). Causal modelling in nonexperimental research. Amsterdam, The Netherlands: Sociometric research foundation.

Saris, W., Den Ronden, J., & Satorra, A., (1987). Testing structural equation models. In P. Cuttance & J. Ecob (Eds.). Structural modelling. Cambridge: Cambridge University Press. (in press)

Saris, W.E., Satorra, A., & Sörbom, D., (1987). The detection and correction of specification errors in structural models. (submitted for publication)

Satorra, A., & Saris, W.E., (1985). The power of the likelihood ratio test in covariance structure analysis. Psychometrika, 50, 83-90.

Sörbom, D., (1987). Model modification. (submitted for publication)

Steiger, J. H., Shapiro, A. & Browne, M., (1985). On the multivariate asymptotic distribution of sequential chi-square statistics. Psychometrika, 50, 253-264.

PITFALLS FOR FORECASTERS

Ton Steerneman and Gertjan Rorijs
University of Groningen   AMRO bank
Institute of Econometrics   Amsterdam, The Netherlands
Groningen, The Netherlands

ABSTRACT

Electronic computers and the implemented statistical software are
of big help to economists and statisticians, e.g. when they are
preparing yearly forecasts of macro-economic quantities. Two important
problems may then occur, namely overfitting and datamining, which have
been considered separately in literature, but the very close relationship between the two does not seem to have been discussed. Trying to
avoid overfitting researchers often delete insignificant variables from
regression equations, and try to find relatively small and appealing
regression equations. This searching process is known as datamining.
The relation between overfitting and datamining will be discussed, and
illustrations will be given of the dangers involved. It will be
concluded that in practice there is no escape possible from both
dangers.

1. INTRODUCTION

The impact of computers on economics and statistics is tremendous.
In the past many tasks of a statistical or a computational nature had to
be done by hardened experts, but at present many of these jobs can
easily be carried out by un-initiated people. The advance of computer
technology has led to a revolutionary growth of the use of statistical
methods in various disciplines. Important features of the computer are
(i) automatic data recording and handling, (ii) the possibility to store
and retrieve an enormous amount of data, and (iii) high computation
speed. A consequence of these enormous computational facilities is that
nowadays anyone can apply standard software without clearly knowing what
he is doing: he thinks the computer will provide the good answers.

With regard to statistical applications it became possible to

supplement elementary analyses by a study of the implications of various
alternative models, including those leading to heavy computations. In
many situations the number of variables is large in comparison to the
sample size. So, with the computational burden largely removed, there is
a real temptation to build models with large numbers of variables. The
danger inherent to such a use of the computer is that statistical
procedures may suffer from a so-called peaking phenomenom: they may show
a decreasing actual performance if the number of variables is increased
beyond a certain bound. This problem was recognized many decades ago.
Reichenbach (1949) called it the problem of the reference class (page
374). Fisher (1938) is another reference and also Rao (1949), who stated
that 'It does not seem to be, always, the more the better ...'.

It frequently happens that one chooses a model, by optimizing the
sample-based performance, without realizing that this performance is
data-dependent and hence uncertain. Here the data are used twice, first
for selecting the variables, and secondly for analyzing the model as if
the variables chosen were predetermined. At this second step the uncertainties involved in the choice of the model or the set of variables are
not taken into account. In fact the sample-reuse assessment of
performance tends to be optimistic.

So a fundamental problem is which variables should be selected and
how the quality can be measured of the ultimate model, taking into
account the uncertainties involved in the model choice. In the literature
many methods of variable selection have been proposed depending on the
purpose one has in mind, and a corresponding criterion, such as the mean
squared prediction error if one aims at regression-based prediction. For
review papers reference can be made to e.g. Hocking (1976), Thompson
(1978), Amemiya (1980) and Judge et al. (1980), chapter 11. One reason
for the lack of unicity of the solution to the selection of variables
problem is the fact that two error sources have to be distinguished.
They will be called Scylla and Charybdis, two figures from Homer's
Odyssey. The imagery is due to Schaafsma and Van Vark (1979). The
terrible vortex Charybdis stands for 'multicollinearity' or 'lack of
performance because of overfitting'. On the other hand the six-headed
dragon Scylla represents the monster of using the data twice: first for
choosing the model and selecting the variables, next for evaluating the
data as if the model and the variables were predetermined.

An outstanding example of Scylla is pretesting: estimate (by OLS) a
linear regression equation and test whether a regression coefficient
does not deviate from zero significantly. If not, omit this variable and
estimate the resulting equation again (by OLS). For reviews on pretest-
estimators we refer to Judge and Bock (1978) and Judge et al. (1980).
More extremely, researchers frequently choose a subset of a set of
explanatory variables that performs best in the sample. For instance,
they choose the best set of 3 explanatory variables out of 10 possible
candidates. This is often called datamining, fishing, etc. Lovell (1983)
makes clear that the significance level of the t-test can be much larger
after searching than the claimed 5%, say. Freedman (1983) illustrates
that it is very probable to find a model with a high F-value, even when
there is in fact no relation at all between the dependent and the
explanatory variables. These are situations where Ulysses' skills seem
to be required. Of course there are also cases for which it does not
matter, because the sample sizes are large (e.g. in pattern
recognition). However, it may also happen that it is wise not to leave
the harbour, because the data is very poor and a shipwreck is bound to
come.

In order to warn against the dangers of overinformation, we
present an example of Charybdis and Scylla in section 2 and 3
respectively. In section 4 we present some theory concerning the peaking
phenomenon in the linear regression model with jointly normal
distributed random variables when prediction is the main aim. Some con-
cluding remarks and suggestions will be given in section 5.

The authors strongly feel that a forecaster should try to build
models of modest complexity, and that this should not be done on the
basis of excessive datamining. Moreover, it will not be possible that
the dangers sketched can be evaded completely: Ulysses avoided
Charybdis, but as a result six men of his crew fell prey to Scylla. Due
to Ulysses' good helmsmanship he and the remaining crew could escape.

2. AN EXAMPLE OF PEAKING

In macro-economic practice, forecasts are often generated by means
of very large macro-econometric models. Regarding the Dutch economy, the
Central Planning Bureau (CPB) provides forecasts based on very large

models: FREIA-KOMPAS for quarterly forecasts and FREIA for yearly ones.
In the sequel, only yearly forecasts are considered. Some of the most
important macro-economic quantities to be predicted are for instance
consumption, inflation, unemployment, investment and wages. The GRECON-
team, associated with the Econometrics Institute of the University of
Groningen aims to make similar yearly predictions for these main macro-
economic variables. The results are, of course, compared with those of
the Central Planning Bureau. In contrast to most other macro-economic
forecasts the GRECON-team also provides estimates for the prediction
standard errors.

The team believes that relatively good yearly predictions can be
obtained by means of a relatively small, linear macro-econometric model.
The actual specification, or equivalently, the selection of the
variables of a single behavioural equation comes into being by trying
out several possibilities. However, there are a number of requirements,
such as

(i)   the variables to be included should be relevant from an economic
       point of view and these should also be statistically significant,
       the sign of the estimated coefficient should also be in line with
       generally accepted economic theory,

(ii)   a modest number of variables should be included in the equation,
       say no more than four,

(iii) the fit of the equation should be reasonable; unfortunately this
       is not always possible, e.g. the fit of the behavioural equation
       for investments (excluding houses) is very poor,

(iv)   the multicollinearity should be modest; this means in practice
       that the absolute value of the correlation coefficient between
       two explanatory variables is less than 0.5.

These demands lead to a small model: the behavioural equations contain
only two or three variables in most cases. We refer to Dietzenbacher
et al. (1986) for the presentation of the GRECON-model.

We first focus upon the danger Charybdis, when univariate
prediction of the growth rate of real consumption (c) is the aim. We
assume a jointly normal distribution of c and 19 explanatory variables,
which are more or less ordered by decreasing economic relevance. This
ordening is based upon a priori considerations and not on the data. Here
we shall not discuss which variables were chosen. We refer to

106

Dietzenbacher et al. (1986) for details and data. Yearly observations
are available over the period 1952-1984. A natural measure for the
quality of a regression equation (including constant term), when
prediction is the main aim, is the mean squared error of prediction to
be discussed in section 4. If $\hat{Y}_0$ is a predictor for Y_0, based on the
data set and a new observation on the explanatory variables, then
$MSEP(\hat{Y}_0) = E(\hat{Y}_0-Y_0)^2$, which depends on unknown parameters and will be
estimated in practice. Its estimator will be denoted by $msep(\hat{Y}_0)$. Since
we assumed that the variables are ordered, the attention is restricted
to the next selections of variables: $\{1\},\{1,2\},\ldots,\{1,2,\ldots,19\}$. The
estimated MSEP for selection $\{1,2,\ldots,p\}$ will be denoted by $msep(p)$. The
results are presented in figure 2.1. If more than 5 variables are

Figure 2.1. The estimated mean squared prediction error as a function
of the number of variables in a sample over the period 1952-1984 for
prediction of the growth rate of consumption.

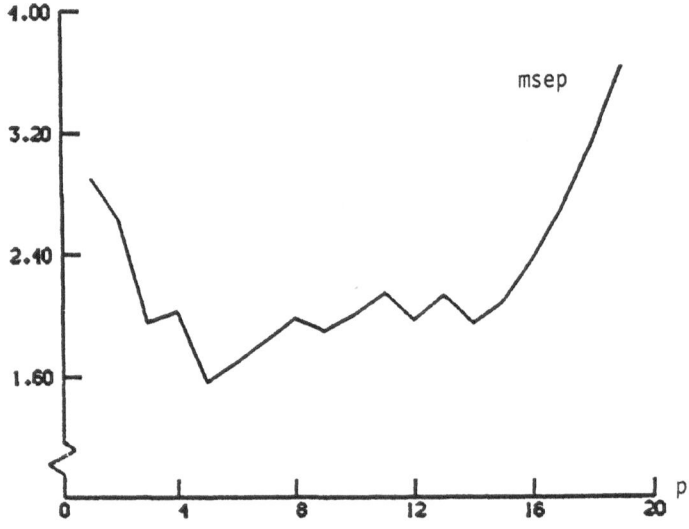

selected, then the performance becomes worse. This illustrates how an
econometrician may observe a peaking phenomenon when he is considering
regression equations which become more complex by adding variables.
    Figure 2.1 suggests that it pays to be parsimonious with the number

of variables in a regression equation, and this way Charybdis can be avoided. The ordering of the variables is of course not compelling. The variables 1 and 2, and possibly 3 are often really important, but the ordering of the remaining variables is less obvious. Now another danger looms up, Scylla. Keeping the model complexity small, it is really tempting to try out a lot of possible combinations of say 2, 3, 4 or 5 variables. In the next section we illustrate that the performance of the regression which is chosen among a lot of possible models looks very nice, but the prediction may be bad.

3. AN EXAMPLE OF DATAMINING

In order to illustrate Scylla we present a simple specification experiment concerning the problem of univariate prediction of the growth rate of consumption (c), investments, excluding houses (i_m), and average wages per year (l). For the regression equations for each of these three variables there were eight possible candidate explanatory variables, the constant term included. All possible regression equations with corresponding performance characteristics were calculated. So, for each of the dependent variables c, i_m and l, we obtained $255 = 2^8 - 1$ regression equations. For the description of the explanatory variables, the dataset, etc. we refer to Dietzenbacher et al. (1986). As a measure for the quality of a regression equation we use again the estimated mean squared prediction error, msep. The msep-criterion explicitly takes into account the aim of prediction. It is usually not applied in econometrics. Here the $\overline{R}^2$- or F-values are common practice. The difference with msep is that $\overline{R}^2$ and F describe some average over-all quality of fit, whereas msep gives an idea of the average prediction error.

We now present the results. The experiments are based on the yearly observations over the period 1952-1983. A prediction is made for the year 1984.

In table 3.1 we also presented minimum and maximum values in order to give an idea of the range of the possible outcomes. If the constant term plays the same part as the 7 other variables, then we may expect the minimum msep-outcome for regressions with 3, 4 or 5 variables,

Table 3.1. Estimated performances of the regression equations
for c, i_m and l.

number of variables	msep c		msep i_m		msep l	
	min	max	min	max	min	max
1	2.58	24.10	47.4	123.9	6.68	96.77
2	1.72	24.90	34.1	122.4	3.75	89.79
3	1.68	18.13	34.5	126.0	3.16	28.42
4	1.52	9.45	35.5	116.7	2.69	13.62
5	1.49	7.61	37.9	99.2	2.45	11.90
6	1.50	4.64	39.9	62.7	2.52	9.04
7	1.62	2.23	43.1	52.6	2.69	3.86
8	1.76	1.76	46.9	46.9	2.92	2.92

because there are 56 = $\binom{8}{3}$, 70 = $\binom{8}{4}$, respectively 56 = $\binom{8}{5}$ possible
regression equations, and it is very probable that there is a low
outcome of msep between these 182 possible regressions. Indeed, for c
and l we find that the 'optimal' number of variables is 5. For i_m we see
that this number equals 2.

By taking the square root of the minimal msep-values we can
conclude from table 3.1 that, on the average, the predictions will
deviate from the realizations with about 1.4 for c, 6.5 for i_m and 1.8
for l. These values are, of course, crude. Forecasting of i_m is always a
nasty affair. This fact is well-known to econometricians involved in
macro-economic model-building.

The 'optimal' regression equation for c contains a constant term.
This is not the case for i_m and l. Incorporating a constant term implies
that there is an autonomous growth in e.g. consumption. This assertion
is not very attractive from an economic point of view. However, this
constant term is not significant at the 5% level. The other 'optimal'
regression equations also contain insignificant contributions. One may
conclude that choosing regression equations with minimal msep is not
sufficiently parsimonious.

The 'optimal' msep value also gives too optimistic a picture,
because we choose the minimum msep from 255 possible values. It is very
probable that a value is chosen which underestimates the mean squared
prediction error in the population.

Constructing predictions for 1984 we use the 'optimal' equations
and the equations as specified in the model GRECON 85-B. The forecasts
are compared with the realizations for 1984.

Table 3.2. Predictions and realizations of c, i_m and l for 1984.

prediction method	outcomes		
	c	i_m	l
minimal msep	0.7	16.4	3.6
specification GRECON 85-B	-0.8	9.7	3.7
realization	-0.5	4.7	1.1

The regression equations as specified by the model GRECON 85-B lead to better forecasts for c and i_m than the 'optimal' equations, and for l the two methods are equally bad. From an economist's point of view the prediction by the 'optimal' method for c is especially bad, because the sign is wrong: a growth of consumption is predicted, while in fact the consumption decreases. The specification requirements of GRECON (see section 2) lead to better results, although the msep outcomes were 1.90 for c, 51.1 for i_m and 3.16 for l. When these results are compared with those of table 3.1, then we see that values obtained for the GRECON-specifications are certainly not optimal. So, the minimal msep outcomes are misleading, if they are used to measure the 'actual' performance of 'optimal' prediction equations (Scylla).

4. SOME THEORETICAL CONSIDERATIONS

In the preceding sections we assumed joint normality of the variable to be predicted (η) and the explanatory variables ($\xi_1,\ldots,\xi_k$). The model we consider can be formulated as follows. Define $\xi = (\xi_1,\ldots,\xi_k)$, then

$$\begin{pmatrix}\eta\\\xi\end{pmatrix} \sim N_{k+1}\left(\begin{pmatrix}\mu\\\nu\end{pmatrix} , \begin{pmatrix}\omega_0^2 & \Omega'\\\Omega & \Sigma\end{pmatrix}\right), \tag{4.1}$$

where $\mu \in \mathbb{R}$ and $\omega_0^2 > 0$. We assume that the $\xi_1,\ldots,\xi_k$ are ordered by decreasing (economic) relevance. Besides the discussion would become hopelessly complicated without this assumption. More intrinsically one may require that the correlation coefficient between η and ξ_{p+1} conditionally given $\xi_1,\ldots,\xi_p$ decreases as p increases. This will be discussed later on. Since we would like to consider models for η based upon $\xi_1, \ldots, \xi_p$ where $p \leq k$, we introduce the notation $\xi^{(p)} = (\xi_1, \ldots, \xi_p)'$. Analogously $\nu^{(p)}$ and $\Omega^{(p)}$ denote the vectors consisting of the first p components of ν and Ω respectively. $\Sigma^{(p)}$ equals the variance-covariance matrix of $\xi^{(p)}$. Note that

$$\begin{pmatrix} \eta \\ \xi^{(p)} \end{pmatrix} \sim N_{p+1} \left(\begin{pmatrix} \mu \\ \nu^{(p)} \end{pmatrix}, \begin{pmatrix} \omega_0^2 & \Omega^{(p)'} \\ \Omega^{(p)} & \Sigma^{(p)} \end{pmatrix} \right), \quad p = 0,1,2,\ldots,k \ . \tag{4.2}$$

From (4.2) we obtain

$$\eta = \alpha(p) + \xi^{(p)'} \beta(p) + \varepsilon(p), \quad p = 0,\ldots,k \tag{4.3}$$

where

$$\beta(p) = \Sigma^{(p)^{-1}} \Omega^{(p)}, \quad \alpha(p) = \mu - \nu^{(p)'} \beta(p)$$

$$\varepsilon(p) \sim N(0,\omega_p^2), \quad \omega_p^2 = \omega_0^2 - \Omega^{(p)'} \Sigma^{(p)^{-1}} \Omega^{(p)},$$

$\xi^{(p)}$ and $\varepsilon(p)$ are independently distributed. In case that we do not want to consider a constant term in the population regression equations (4.3) we take $\mu = 0$ and $\nu = 0$ in (4.1).

If the sample size n increases, one will be willing to enter more variables into the analysis. The maximum number of variables $k = k(n)$ will increase, and we will also have $k(n) \to \infty$ as $n \to \infty$. So it is interesting to see what happens if it is allowed that $k \to \infty$ in (4.1). Since ω_p^2, the variance of the disturbance term in (4.2) is decreasing in p, we may define $\omega_\infty^2 = \lim \omega_p^2$. If $\omega_\infty^2 > 0$, then there is some intrinsic noise in   the system. In case $\omega_\infty^2 = 0$, then a completely deterministic model describing η will be obtained as $p \to \infty$.

In practice the parameters will be unknown. Assume that we have at our disposal an independent sample $\begin{pmatrix} Y_1 \\ X_1 \end{pmatrix}, \ldots, \begin{pmatrix} Y_n \\ X_n \end{pmatrix}$ and a realization X_0 of ξ. We have to predict Y_0. Taking p variables into account, (4.3) is estimated by OLS on basis of the sample. Two cases have to be distinguished, because the constant term may or may not be involved in (4.3). $\hat{Y}_{0p}$ will denote the usual classical predictor of Y_0 obtained by substituting $X_0^{(p)}$ in the estimated regression equation, ignoring disturbances. A natural measure of the quality or performance of $\hat{Y}_{0p}$ is the so-called mean-squared error of prediction: $MSEP(n,p) = E(Y_0 - \hat{Y}_{0p})^2$.

Relatively tedious computations are involved in deriving an explicit
expression, because the expectation runs over the sample and also over
Y_0 and X_0. We have

$$MSEP(n,p) = \begin{cases} (n-1)(n-p-1)^{-1}\omega_p^2 & \text{, regression without constant term,} \\ n^{-1}(n+1)(n+2)(n-p-2)^{-1}\omega_p^2 & \text{, regression including constant term.} \end{cases}$$

For derivations of these formulas we refer to Stein (1960), formula
(2.10), (2.11) and (2.23), Oliker (1978) and for a simpler derivation
Steerneman (1984a). The case of regressions without constant term has
been considered by Breiman and Freedman (1983).

Note that $(n-p-1)^{-1}$ and $(n-p-2)^{-1}$ are increasing with p and that ω_p^2
decreases. Hence $MSEP(n,p)$ will achieve its minimum for some
$p^* \in \{0,\ldots,k\}$, where $k \leq n-3$. In most practical cases we have that p^*
is much smaller than k. In the sequel we assume that p^* is chosen such
that p^* is the smallest value of p for which $MSEP(n,p)$ is minimized.
There exists a so-called optimum measurement or model complexity. In
practice p^* will be relatively small in comparison to n. So, it will be
worse to introduce too many variables. This fact has been described as
the curse of dimensionality in the literature about statistical pattern
recognition. It is interesting to see what happens if $n \to \infty$ and
$k(n) \to \infty$. One may think of $k(n) = n-3$ or $k(n) = \frac{1}{2}n$ for example. Of
special interest is the behaviour of $p^*(n)/n$ as $n \to \infty$. This has been
studied by Breiman and Freedman (1983) (in case $\omega_\infty^2 > 0$) and Steerneman
(1984a,b) ($\omega_\infty^2 = 0$ and $\omega_\infty^2 > 0$). The results can be summarized in the
following theorem.

Theorem 4.1. Consider regression including constant term.
(a) If $\omega_\infty^2 > 0$, then $p^*(n) \to \infty$ and $p^*(n)/n \to 0$.
(b) If the sequence $\{\omega_{p+1}^2/\omega_p^2\}$ is nondecreasing then $MSEP(n,p)$ is
    decreasing for $p \leq p^*$ and increasing for $p \geq p^* + 1$.
(c) Let $k(n) = n-3$. If the sequence $\{\omega_{p+1}^2/\omega_p^2\}$ is nondecreasing then
    $\omega_{p+1}^2/\omega_p^2 \to 1$ if and only if $n - p^*(n) \to \infty$.
(d) Let $k(n) = n-3$. If the sequence $\{\omega_{p+1}^2/\omega_p^2\}$ is nondecreasing and

$\omega_{p+1}^2/\omega_p^2 \to 1$, then $\delta \in [0,\infty]$ is a limit point of the sequence $\{p(1-\omega_{p+1}^2/\omega_p^2)\}$ if and only if $\delta/(1+\delta)$ is a limit point of p^*/n.

Regarding the conditions mentioned in the theorem one should note that if $\omega_\infty^2 > 0$, then automatically $\omega_{p+1}^2/\omega_p^2 \to 1$. Note that $\{\omega_{p+1}^2/\omega_p^2\}$ is non-decreasing if and only if $\{(\omega_p^2-\omega_{p+1}^2)/\omega_{p+1}^2\}$ is nonincreasing. Under this condition the variables are ordered by nonincreasing conditional correlation coefficients, because $[(\omega_p^2-\omega_{p+1}^2)/\omega_{p+1}^2]^{\frac{1}{2}}$ is equal to the conditional correlation between η and ξ_{p+1} given $\xi_1,\ldots,\xi_p$.

We now present some examples.

Example 4.1. We shall consider a model aiming to predict η on the basis of explanatory variables $\xi_1,\xi_2,\ldots$, which are equally relevant. Let η be a latent variable. Assume that $\xi_i = \rho\eta + U_i$, $\eta,U_1,U_2,\ldots$ are independently distributed with $\eta \sim N(0,\sigma_\eta^2)$, $U_i \sim N(0,\sigma_u^2)$. Hence

$$
\begin{pmatrix} \eta \\ \xi_1 \\ \vdots \\ \xi_p \end{pmatrix} \sim N_{p+1}\left(\begin{pmatrix} 0 \\ 0 \\ \vdots \\ 0 \end{pmatrix}, \begin{pmatrix} \sigma_\eta^2 & \rho\sigma_\eta^2 & \cdots & \rho\sigma_\eta^2 \\ \rho\sigma_\eta^2 & \rho^2\sigma_\eta^2+\sigma_u^2 & \cdots & \rho^2\sigma_\eta^2 \\ \vdots & \vdots & \ddots & \vdots \\ \rho\sigma_\eta^2 & \rho^2\sigma_\eta^2 & \cdots & \rho^2\sigma_\eta^2+\sigma_u^2 \end{pmatrix} \right).
$$

It can be shown that

$$
\omega_p^2 = \sigma_u^2\sigma_\eta^2(\sigma_u^2+p\rho^2\sigma_\eta^2)^{-1} .
$$

If we take $k(n) = n-3$, then it follows from theorem 4.1 (d) that $p^*/n \to \frac{1}{2}$. So, in the case of complete symmetry regarding the influence of the explanatory variables on the dependent variable and concerning the multicollinearity between the explanatory variables asymptotically the number of variables to be selected is approximately $\frac{1}{2}n$. Note that if the parameters were known we could achieve perfect prediction if all variables are selected since $\omega_\infty^2 = 0$ in this example. For we do not know the parameters it is wise to restrict the number of variables to be chosen.

Example 4.2. Consider the model

$$\eta = \sum_{i=1}^{\infty} i^{-1} \xi_i + \varepsilon,$$

where $\varepsilon, \xi_1, \xi_2, \ldots$ are independently distributed, $\varepsilon \sim N(0, \sigma_\varepsilon^2)$, $\xi_i \sim N(0, \sigma_\xi^2)$. In this context the influence of ξ_i on η decreases as i increases. Note that $\sum_{i=1}^{\infty} i^{-2} = \pi^2/6$, then we see that var $\eta = \sigma_\varepsilon^2 + \sigma_\xi^2 \pi^2/6$. It can be derived in an easy way that

$$\omega_p^2 = \sigma_\varepsilon^2 + \sigma_\xi^2 \sum_{i=p+1}^{\infty} i^{-2} .$$

It follows that $p^*/n \to 0$ in case $\sigma_\varepsilon^2 > 0$. When $\sigma_\varepsilon^2 = 0$, then $p^*/n \to \frac{1}{2}$. In case of known parameters and $\sigma_\varepsilon^2 > 0$ we cannot have perfect prediction if all variables are selected, because $\omega_\infty^2 = \sigma_\varepsilon^2 > 0$. When the parameters are unknown and have to be estimated one should be very parsimonious with the choice of the number of variables: the optimal number of variables (p^*) is very small relatively to the sample size, because $p^*/n \to 0$. If $\sigma_\varepsilon^2 = 0$ one can be more daring: $p^*/n \to \frac{1}{2}$.

It is clear that it is advantageous to know p^*, in practice this is of course impossible because we do not know MSEP(n,p). The curve of MSEP can be estimated by replacing ω_p^2 by its UMVU estimator $\hat{\omega}_p^2$. So we obtain the estimator msep(n,p). In the literature it has been proposed to estimate p^* by $\hat{p}$, which is the smallest possible solution to the problem minimizing msep(n,p), which is also well-known as the S_p criterion. We refer to Hocking (1976) and Thompson (1978). In case that there are no intercepts in the regressions Breiman and Freedman (1983) showed that $\hat{p}$ is asymptotically optimal in the sense that

$$\text{msep}(n,\hat{p})/\text{MSEP}(n,p^*) \to 1$$

in probability as $n \to \infty$. Steerneman (1984a,b) established that this property is not very compelling, because there exist a lot of other selection-of-variables procedures which are also asymptotically optimal.

The choice of the model complexity $\hat{p}$ is based upon the estimated curve msep(n,p), and it may be worthwhile to construct confidence bounds.

In the following simulation study we try to sketch some peculiar phenomena. Consider the model

$$\eta = \alpha + \sum_{i=1}^{\infty} \beta_i \xi_i + \epsilon,$$

where

$$\alpha = 1, \ \beta_i = 2^{3-i}, \ i = 1,2,\dots$$

$\alpha, \xi_1, \xi_2, \dots$ are independent standard normally distributed.

It can easily be derived that $\omega_p^2 = \frac{1}{3}4^{3-p} + 1$. In our simulation experiment we take n = 35 and we shall consider k = 30 variables. The following quantities can be calculated in the simulation: MSEP(n,p); msep(n,p), the UMVU-estimate based on the sample; the standard deviation of msep(n,p), where p = 0,1,2,...,30. Note that $(n-p-1)\hat{\omega}_p^2/\omega_p^2$ follows a χ^2-distribution with n-p-1 degrees of freedom, hence

$$\mathrm{var}(\mathrm{msep}(n,p)) = 2(n-p-1)^{-1}(\mathrm{MSEP}(n,p))^2.$$

The results are presented in figure 4.1.

<u>Figure</u> 4.1. On estimating the curve of MSEP(n,p) in the simulation study.

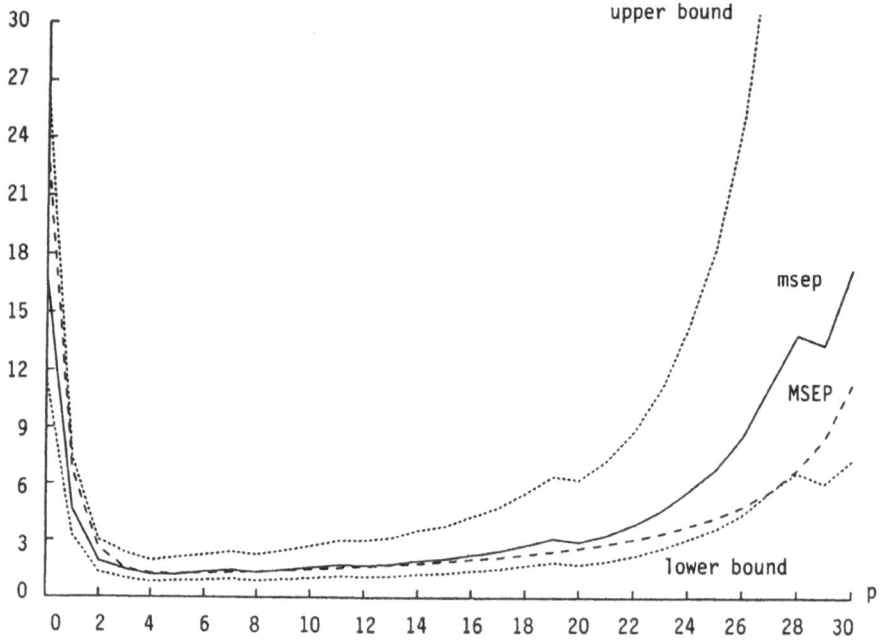

Figure 4.1 shows a 'peaking' phenomenon for MSEP, and $p^* = 5$, which is considerably smaller than 30. So, it pays to delete variables, although

it is not very disadvantageous to take p = 10 for instance, but it does
not help to get better predictions. In this particular case one observes
that MSEP varies only slightly for quite a large range of p. Note that
MSEP(35,33) = ∞. The curve of MSEP becomes rapidly increasing for p
larger than 30. In practice, of course, we do not know MSEP and we apply
its UMVU-estimator. From figure 4.1 we observe that it seems as if
msep(n,p) is a better estimator for MSEP around $p = p^*$ than elsewhere.
In fact the standard deviation seems to possess a 'peaking property'
too; this can easily be checked from figure 4.1. It appears that msep is
a good estimator for MSEP only around $p = p^*$. For $p < p^*$ we have underestimation and for $p > p^*$ we see that there is overestimation. We do not
have a general explanation for this phenomenon yet. On the whole,
msep(n,p) does not seem to be a very good estimator for MSEP(n,p), which
is certainly disconcerting when we try to estimate p^* by $\hat{p}$. Therefore,
it might be useful to have confidence limits for MSEP(n,p) p = 0,...,k,
which can easily be obtained from the χ^2-distribution, but for larger
values of p, the 90% confidence intervals are rather asymmetric. Further
research is needed for obtaining smaller confidence intervals, the
classical approach is not very satisfactory within this framework.

5. CONCLUDING REMARKS AND SUGGESTIONS

In the previous sections the consequences of overfitting and datamining were illustrated. We do not have a cut-and-dried procedure for
avoiding these two pitfalls. It can be suggested that one can avoid
Charybdis by being parsimonious in choosing the number of variables or
the complexity of the model. However, one will always be victimized by
Scylla, because researchers will always have to choose between several
possible models, but maybe the damage can be restricted.

What should be done is to keep the alternative models small.
Besides the number of alternative models should be drastically
restricted on the basis of a priori information. A researcher should
contact economists or other experts and he should try to develop some
theory (not using the data) or some alternative theories. So, do not
start using the computer too early. When finally the computer is needed
always check the whole output, not only the figures you were looking

for. Before starting the statistical analysis a lot of combinations of variables can be eliminated, and a certain vague ordering of the variables can be proposed. These considerations may lead to a relatively small number of possible models, which enter the data-mining process.

ACKNOWLEDGEMENTS

The authors would like to thank Prof.Dr. W. Schaafsma, Prof.Dr. T.J. Wansbeek, and Dr. T.K. Dijkstra for their helpful comments.

REFERENCES

Amemiya, T. (1980), Selection of regressors; International Economic Review, vol. 21, 331-354.
Breiman, L. and D. Freedman (1983), How many variables should be entered in a regression equation?; Journal of the American Statistical Association, vol. 78, 131-136.
Dietzenbacher, H.W.A., A.G.M. Steerneman, V.J. de Jong, M.A. Kooyman and W. Voorhoeve (1986), Het model GRECON 86-B en de voorspellingen voor 1986; Rapport 86-06-SE, Econometrisch Instituut, Rijksuniversiteit Groningen.
Fisher, R.A. (1938), On the statistical treatment of the relation between sea level-characteristics and high-altitude acclimatization; Proceedings of the Royal Society of London, ser. B, vol. 126, 25-29.
Freedman, D.A. (1983), A note on screening regression equations; The American Statistician, vol. 37, 152-155.
Hocking, R.R. (1976), The analysis and selection of variables in linear regression; Biometrics, vol. 32, 1-49.
Judge, G.G. and M.E. Bock (1978), The statistical implications of pre-test and Stein-rule estimators in econometrics; North-Holland Publishing Company.
Judge, G.G., W.E. Griffiths, R.C. Hill, T.-C. Lee (1980), The theory and practice of econometrics; John Wiley & Sons, Inc.
Lovell, M.C. (1983), Data Mining; The Review of Economics and Statistics, vol. 65, 1-12.
Oliker, V.I. (1978), On the relationship between the sample size and the number of variables in a linear regression model; Communications in Statistics, Theory and Methods, vol. A7, 509-516.
Rao, C.R. (1949), On some problems arising out of discrimination with multiple characters; Sankhyā, vol. 9, 343-364.
Reichenbach, H. (1949), The theory of probability; University of California Press, Berkeley and Los Angeles.
Schaafsma, W. and G.N. van Vark (1979), Classification and discrimination problems with applications IIa; Statistica Neerlandica, vol. 33, 91-126.
Steerneman, A.G.M. (1984a), Prediction performance and the number of variables in multivariate linear regression; theorems and proofs; Rapport 84-01-SE, Econometrisch Instituut, Rijksuniversiteit Groningen.

Steerneman, A.G.M. (1984b), Prediction performance and the number of
    variables in multivariate linear regression; pages 118-129 of
    Dijkstra, T.K., editor (1984), Misspecification Analysis; Lecture
    Notes in Economics and Mathematical Systems 237, Springer-Verlag.
Stein, Ch. (1960), Multiple regression; ch. 37 from Olkin, I.,
    S.G. Ghurye, W. Hoeffding, W.G. Madow, H.B. Mann, editors (1960),
    Contributions to Probability and Statistics (Essays in Honor of
    Harold Hotelling), Stanford University Press, Stanford California.
Thompson, M.L. (1978), Selection of variables in multiple regression;
    International Statistical Review, vol. 46, 1-9 and 129-146.

MODEL SELECTION IN MULTINOMIAL EXPERIMENTS

JAN DE LEEUW
DEPARTMENT OF DATA THEORY FSW
UNIVERSITY OF LEIDEN
MIDDELSTEGRACHT 4
2312 TW LEIDEN, THE NETHERLANDS

Multinomial experiments and models

We first define what we mean by a multinomial experiment. The first component of the definition is a set $\mathcal{P}$, the *population*. Elements of $\mathcal{P}$ are called *objects*. We shall assume in this paper that $\mathcal{P}$ is finite, and that there is a list of $\mathcal{N}$ objects defining this population. The objects are of m different types. Suppose there are $\mathcal{N}_j$ objects of type j, and define the *theoretical proportions* $\pi_j = \mathcal{N}_j/\mathcal{N}$. Now consider the set of all sequences of length N, with elements from $\mathcal{P}$. There are $\mathcal{N}^N$ such sequences, but if types are indistinguishable not all of them are different. In particular there are $\Pi_{j=1}^{m} \mathcal{N}_j^{n_j}$ indistinguishable sequences with *frequencies* n_j, and if the order of the objects in the sequence is irrelevant there are even N! $\Pi_{j=1}^{m} \{\mathcal{N}_j^{n_j}/n_j!\}$ such sequences. The proportion of sequences with given frequencies, where order is not taken into account, is consequently $\text{prop}(n_1,...,n_m) = N! \ \Pi_{j=1}^{m} \{\pi_j^{n_j}/n_j!\}$. With each sequence of length N we associate the *vector of proportions* p, with $p_j = n_j/N$, and the *multinomial probability* $\text{prop}(n_1,...,n_m)$. Another way of expressing this is that for each N we have defined a random vector $\underline{p}_N$, taking values in S^{m-1}, the unit simplex in $\mathcal{R}^m$. We use the convention of underlining random variables in this paper. Thus $\underline{p}_N$ is a sequence of random vectors.

This describes the multinomial experiment. We have the pair $(\mathcal{N},\pi)$ and the sequence $\underline{p}_N$, and generally $(\mathcal{N},\pi)$ is unknown. The first and most basic problem we shall study is the *estimation* of π. This means, informally, that we construct a new sequence of random variables $\Phi(\underline{p}_N)$ which is (in some sense) as close as possible to the unknown π. Observe that we restrict our attention to functions of the proportions $\underline{p}_N$. Precise definitions of closeness and optimality will be given below.

In order to make our estimates accurate, we must try to take as much prior information into account as possible. Science is cumulative, and presumably we already know something about the subject area in question. Prior information takes the form of a *model* in this paper. A model Ω is a subset of S^{m-1}. If we say that model Ω is *true*, then we mean that $\pi \in \Omega$. The second problem we shall investigate in this paper, if assuming that a model is true will help us to compute more precise estimates, even in those cases in which the model actually is only approximately true. This could be formulated as deciding whether a model is *useful.* Moreover, and finally, we shall study a similar problem for the case in which we have a finite number of models $\Omega_1,...,\Omega_t$. This is one version of the problem of *model choice*, in which we want to find out which model helps us best to improve our estimates, i.e. which model is most useful.

Up to now we have defined a mathematical structure called a multinomial experiment, but we have not yet specified how this structure is connected with empirical data. Suppose we have selected N objects from the population. We can compute the m quantities n_j, which are the *observed frequencies* of objects with type j, and $p_j = n_j/N$, which are the *observed proportions*. We assume now, that p is a *realization* of the random variable $\underline{p}_N$. This means that we compare statistics computed from p with statistics computed from the other possible realizations of this random variable. This defines our *framework of replication* (De Leeuw, 1984), the set of all possible outcomes with which we compare our results. Because our framework gives all sequences the same probability this means that we act as if that our sample is a simple random sample, drawn with replacement from a finite population. It is important to realize that all statistical statements are about this framework $\underline{p}_N$, not about the data p, and also not about the true value π. It is also important to see that a framework is never true or untrue (for a particular empirical situation), but it is either relevant or irrelevant. We do not assume, in any sense of the word, that our sample is indeed a random sample. Our results have a simple combinatorial interpretation. The counting of samples, familiar from combinatorics, is done by using asymptotic approximations. Thus the 'foundational' and 'inferential' aspects of statistics, which are both controversial and problematical, are not relevant for our discussion. We go 'back to the Laplace definition' (Hemelrijk, 1968).

It is perhaps worth mentioning that our results, which are true for multinomial situations, can be generalized with little effort to product-multinomial situations in which we deal with more than one population (or with a stratified population). With a bit more effort they can be extended to more complicated sampling designs. And with a considerable amount of technical effort they can also be extended to infinite dimensional problems (functions of empirical distribution functions, regression functions, or density estimates). The basic methodological ideas and interpretations remain the same in these alternative situations.

Models are never true

There has recently been much discussion in statistics about the role of models. Compare McCullagh and Nelder (1983, section 1.1), De Leeuw (1984), or Nelder (1984) for fairly modern introductions. The general consensus seems to be that, contrary to the practice of classical statistics, we must not routinely assume that our models are true, i.e. that $\pi \varepsilon \Omega$. Models are approximations, which are summaries of the prior scientific information we have, but which can still be quite far off the mark in some cases. We need models to increase the stability of our estimators, descriptors, and predictors. Using a model which is perhaps not true, but based on only a few parameters, means that we trade statistical stability for unbiasedness. Making too few assumptions means instability, and may not be very cumulative from the scientific point of view. Too many assumptions means a great deal of stability, but possibly a very large bias. We need to find a compromise between the two, and for such a compromise models are necessary. The fact that models are approximations, and that approximation errors may be more important than sampling errors, is stressed by various schools of data analysis. We mention Tukey (1980), Guttman (1985), Benzécri et al. (1973), Gifi (1981), Box (1979), Verbeek (1984), Kalman (1983), Willems (1986).

If we compare the description of multinomial experiments above, for example, with actual data collection or experimentation, then the description involves various *idealizations*. In the first place populations are usually not given by finite lists. The current population of the Netherlands, for instance, is not exactly well defined. We must decide whether we mean all human beings currently within our borders, or all human beings with dutch citizenship, or all human beings registered by the proper authorities as living in Netherlands. It is clear that a little bit of creative thinking shows that there are various borderline cases and exceptions, which make it difficult to construct lists, even in theory. And, more seriously, these populations are changing every day because of births, deaths, naturalizations, immigrations, tourism, and so on.

The second idealization is the idea of sampling with replacement. Of course this can be carried out formally only if we actually have a list, while we have just decided that such lists do not exist. The actual use of 'with replacement' is very rare, except in artificial sampling experiments. In many cases *hypergeometric experiments*, which do not use replacement, are more realistic. Fortunately most of what we say applies directly to hypergeometric experiments as well, and if N is large the difference between the two random variables is small anyway. In actual experiments with human subjects simple comparison with all possible subsets of size N is often not the most interesting comparison, because stratification and clustering are the rule rather than the exception. This means, in other words, than simple random sampling (with or without replacement) does not provide the most relevant replication framework.

And, as we have already seen, our notion of a model itself is an idealization. Models are never true, they are at best good approximations, where 'good' means good enough for practical purposes. Classical mechanics is not true, it is an approximation which is good enough for most purposes. Relativity theory improves the approximation, but this still does not make it true. There is no reason to suppose that we shall ever find a model which is 'true', in the sense that it gives perfect predictions of all phenomena under consideration, or even predictions which can never be improved any more. This notion of truth is not really needed, if only because of the omnipresence of measurement error. That 'truth' exists is, in itself, a model, which may be useful for guiding our actions, but which is not part of science.

The fact that even the simple multinomial model involves many idealizations and approximations which are, at least in many cases, not really appropriate, need not disturb us greatly. The model is meant as an approximation, and this is not only true for the model Ω but also for the framework modelling the sampling procedure. The question is whether these idealizations make it possible to give descriptions and predictions which are useful and good enough for practical purposes, or which are as good as possible under the circumstances. There are many situations in the social sciences in which the idea of random sampling from a well-defined population is much more far fetched than in the simple demographic or survey situations we have in mind here. In order to apply basic statistical ideas much more far reaching idealizations, such as super-populations or infinite hypothetical populations, are needed. One can seriously wonder in many of these cases if the classical statistical approach is really fruitful, although not many systematic alternatives have been proposed, at least for prediction. In such cases a satisfactory description of the data is perhaps all that can be realistically expected. Techniques that try to go beyond mere description must make many assumptions that are often untested or even untestable. This entails that the conclusions in these cases are based largely on prejudice.

Example: twins

In the first part of the paper we use a simple multinomial example, taken from Andersen (1980). Suppose that we study the gender of twins. There are three outcomes: girl-girl, boy-boy, and girl-boy. We start with simple binomial models for the probability of a girl and the probability of a boy. Thus we ignore the fact that twins come in pairs, and we only count the sexes. The saturated binomial model leaves $\text{prob}(\female)$ and $\text{prob}(\male) = 1 - \text{prob}(\female)$ free, the restricted binomial model sets $\text{prob}(\female) = \text{prob}(\male) = 1/2$.

The situation becomes a bit more complicated if we study the multinomial experiment with the three possible combinations of boy and girl as outcomes. The first model one may think of, is that of independence (model A in the sequel). If the probability of a girl is ω, then

$$\text{prob}(\female\female) = \omega^2, \tag{1a}$$

$$\text{prob}(\male\male) = (1 - \omega)^2, \tag{1b}$$

$$\text{prob}(\female\male) = 2\omega(1 - \omega). \tag{1c}$$

Is this a useful model ? In order to study this question we first need data. In Figure 1 we have drawn the simplex S^2, and the one-dimensional quadratic manifold defined by model A. We have also drawn in the data points for samples from six birth centres. These data are given in Table 1, which is taken from Andersen (1980, page 93).

	2 boys	2 girls	1 boy, 1 girl
Melbourne	29	36	33
Sao Paulo	61	69	81
Santiago	88	77	76
Alexandria	116	114	161
Hong Kong	45	46	34
Zagreb	20	32	30

Table 1: twin data

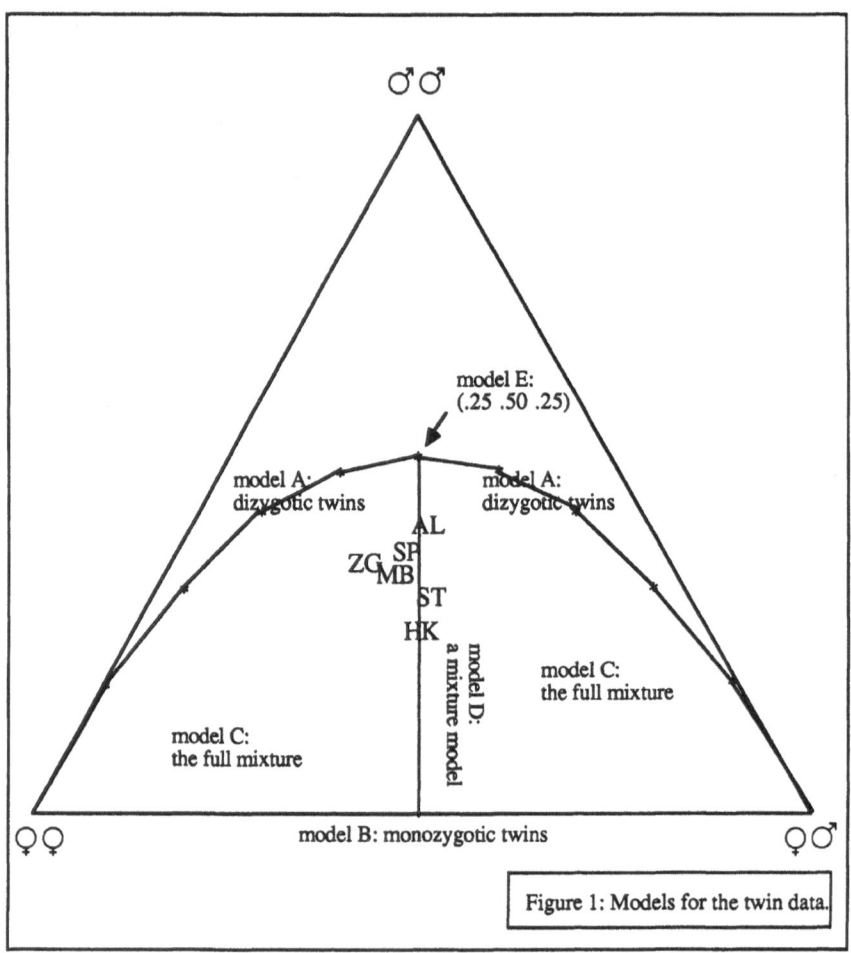

Figure 1: Models for the twin data.

It seems clear from the data that there are fewer girl-boy twins than expected on the basis of independence. The explanation is that twins are either monozygotic or dizygotic, and that monozygotic twins are always of the same sex. For monozygotic twins only we consequently hav

$$\text{prob}(\female\female) = \theta, \qquad (2a)$$
$$\text{prob}(\male\male) = 1 - \theta, \qquad (2b)$$
$$\text{prob}(\female\male) = 0. \qquad (2c)$$

This is, of course, the basis of the triangle in Figure 1. Model B fits badly, which is not very surprising because certainly not all twins are monozygotic.

What we really need is a mixture of models A and B, in which we suppose that the proportion of monozygotic twins is λ, an additional parameter. This mixture model C is the convex region between the 'monozygotic' basis of the triangle and the 'dizygotic' quadratic of model A, and clearly all data points are in this region. A more specific mixture model D supposes that $\omega = \theta = .5$. Then

$$\text{prob}(\female\female) = (1 + \lambda)/4, \qquad (3a)$$
$$\text{prob}(\male\male) = (1 + \lambda)/4, \qquad (3b)$$
$$\text{prob}(\female\male) = (1 - \lambda)/2. \qquad (3c)$$

This is the vertical line segment in Figure 1, which seems to describe the data quite well.

Models A, B, and D are all one-dimensional. Model C is two-dimensional, but still restrictive. For the sake of completeness we also define the zero-dimensional model E, which is the intersection of A and D. It consists of the single point (0.25 0.50 0.25). The question is if these models help us to find better estimates of π From the figure it would seem that model D is best, but we would like to have one or more procedures that make it possible to make a choice if the situation is less clear. For this we need some formal statistical theory, which we introduce in a somewhat unconventional way. The methods we use are inspired by the treatment of multinomial experiments in the book of Rao (1973), but also by the recent emphasis on geometrical methods in statistical estimation theory, and by the equally recent work on resampling methods.

Minimum distance methods

An *estimator* for a multinomial experiment is a continuous mapping Φ of S^{m-1} into S^{m-1}. The idea is, that we try to estimate the population value or true value π. We do this by associating an *estimate* $\Phi(p)$ with each vector of observed frequencies p. The estimator Φ can be thought of as a random variable, the estimate $\Phi(p)$ is a realization of this variable. The identity mapping is an obvious estimator, but in some circumstances it may be possible to improve on this estimator by taking prior information into account. In our case the prior information takes the form of a model Ω, in other cases, which we do not study here, it may take the form of a prior distribution. The procedures we develop have a frequency interpretation even if they use prior distributions, because our replication framework simply remains the counting of samples.

The problem we study is how to estimate π 'optimally'. For this we first have to define optimality, of course. The theory we shall explain is more limited in scope than other forms of statistical large sample theory, but sufficiently general to cover many situations of practical interest. It uses the currently popular geometrical terminology of distance, projection, and manifolds. Introduce a *multinomial separator*, which is a function Δ on $S^{m-1} \times S^{m-1}$ with the properties that $\Delta(p,q) \geq 0$ for all p,q, and $\Delta(p,q) = 0$ if and only if p = q. Throughout the paper we shall not be concerned with regularity conditions. We simply assume that Δ is sufficiently many times differentiable for our results to be true. Given the separator Δ, and the model Ω, we now define the estimator Φ by $\Phi(p) = \text{argmin } \{\Delta(p,q) \mid q \in \Omega\}$, where we assume that for each p ε S^{m-1} the minimum exists, and is unique. Thus $\Phi(p)$ is the 'Δ-projection' of p on Ω. Compare Figure 2.

It follows directly that $\Phi(p) = p$ for all p ε Ω. This condition is known as *F-consistency*, where the F stands for Fisher, who introduced the concept in the twenties. If Φ is F-consistent for a given model Ω containing π, and Φ is also differentiable, it follows that the distribution of $N^{1/2}(\Phi(\underline{p}) - \pi)$ converges to a multivariate normal distribution. We say that Φ is CAN, or *consistent asymptotically normal*.

Our comparison of models will be based on the following two statistics. Compare Figure 2. In the first place $\Delta(\pi,\Phi(\underline{p}))$, the *estimation error*, i.e. the distance from $\Phi(\underline{p})$ to the true value π. It is obvious how the estmation error must be interpreted. It tells us how far off we are if the observed value is $\underline{p}$ and the true value is π. In the second place we study the *prediction error* $\Delta(\underline{q},\Phi(\underline{p}))$, with $\underline{q}$ an independent vector of proportions from the same multinomial distribution. This distance shows how well the estimate $\Phi(\underline{p})$ predicts an independent replication. It is used to *cross validate* the estimate. Observe that both the estimation error and the prediction error are in general nondegenerate random variables, and that we consequently study their distribution.

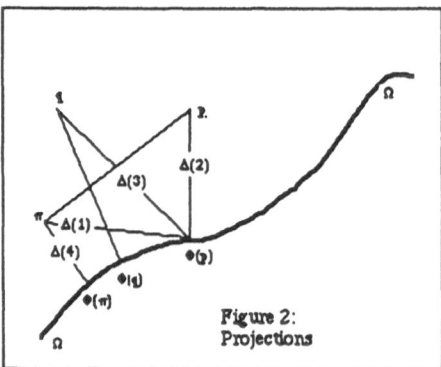

Figure 2: Projections

The third variable we use, among other things to compute our estimates, is the *projection distance* $\Delta(\underline{p},\Phi(\underline{p})) = \min \{\Delta(\underline{p},q) \mid q \in \Omega\}$. We are not really interested in the projection distance as such, but it is a very important intermediate quantity. The reasons for this are clear. In the usual cases in data analysis we do not have a realization available of $\Delta(\pi,\Phi(p))$, because π is unknown, and of $\Delta(q,\Phi(p))$, because we do not have a second independent sample. We only have a single realization of $\Delta(p,\Phi(p))$.

It is instructive to find out what happens with the statistics we have introduced in the 'extreme' cases $T\Omega = S^{m-1}$ and $\Omega = \{\pi_0\}$, where π_0 is not necessarily equal to the true value π. For the zero dimensional model π_0 the estimation error is nonrandom, and equal to *approximation error* $\Delta(\pi,\Phi(\pi))$, while the prediction error and the projection distance are two independent copies $\Delta(p,\pi_0)$ and $\Delta(q,\pi_0)$ of the same random variable. For the *saturated model* S^{m-1} we have a zero projection distance, and a zero approximation error. The estimation error is $\Delta(\pi,p)$ and the prediction error is $\Delta(q,p)$. We shall now proceed to compute asymptotic distributions of these separator-statistics, emphasizing the limit of their expectations. The expected value of the estimation error is called the *bias*, the expected value of the prediction error we call the *distortion*.

The first result is very simple. Clearly the expected values of estimation error, prediction error, and projection distance are asymptotically equal to the approximation error $\Delta(\pi,\Phi(\pi))$. Thus $E(\Delta(p,\Phi(p))) = \Delta(\pi,\Phi(\pi)) + o(1)$, and we have the same expression for $E(\Delta(\pi,\Phi(p)))$ and $E(\Delta(q,\Phi(p)))$. This means that we can estimate the approximation error, the bias, and the distortion consistently by using the projection distance. This result is not very satisfactory, however, because the projection distance is, by definition, smallest for high-dimensional models, and is zero for $\Omega = S^{m-1}$. Thus 'large' models always give small estimates of bias and distortion, the larger the model the smaller the estimate. This is contrary to the well-known phenomenon that prediction error increases if our models are too large. If we want to obtain more precise comparisons, we need more precise approximations.

Let us introduce the following short-hand notation for the partial derivatives. We use $\lambda(p,q)$ for $D_1\Delta(p,q)$, and $\eta(p,q)$ for $D_2\Delta(p,q)$. For the second derivatives we use $A(p,q) = D_{11}\Delta(p,q)$, $B(p,q) = D_{22}\Delta(p,q)$, and $C(p,q) = D_{12}\Delta(p,q)$. If the arguments are not indicated the derivatives are evaluated at $(\pi,\Phi(\pi))$, which is equal to (π,π) if $\pi \ \varepsilon \ \Omega$. For $D\Phi(p)$ we use $G(p)$, G without argument is evaluated at $\Phi(\pi)$. The matrix $T_j(p)$ contains the second partials $D^{(2)}\phi_j(p)$. With T_j we again mean $T_j(\Phi(\pi))$. Finally we define $\Gamma = \Sigma_{j=1}^m \eta_j T_j$. It is not difficult to see that the properties assumed for Δ imply that $\lambda(p,p) = \kappa u$ fo some κ, where u has all elements equal to +1, and $\eta(p,p) = -\kappa u$. Moreover $B(p,p) = A(p,p) = -C(p,p) = -C'(p,p)$ for all p in S^{m-1}. These four matrices with second order partials are all positive semidefinite.

Now let $\underline{\delta} = N^{1/2}(p - \pi)$. Then $\underline{\delta}$ is asymptotically normal with mean zero, and variance $V = \Pi - \pi\pi'$, $\Pi = \text{diag}(\pi)$. We write

$$\Delta(\pi,\Phi(p)) = \Delta(\pi,\Phi(\pi + N^{-1/2}\underline{\delta})) = \Delta(\pi,\Phi(\pi)) + N^{-1/2}\eta'G\underline{\delta} +$$
$$+ 1/2 \ N^{-1}\underline{\delta}'\{\Gamma + G'BG\}\underline{\delta} + o_p(N^{-1}). \tag{4a}$$

In the same way

$$\Delta(\underline{p},\Phi(\underline{p})) = \Delta(\pi + N^{-1/2}\underline{\delta},\Phi(\pi + N^{-1/2}\underline{\delta})) = \Delta(\pi,\Phi(\pi)) + N^{-1/2}(\lambda'\underline{\delta} + \eta'G\underline{\delta}) +$$
$$+ \ 1/2 \ N^{-1}\underline{\delta}'\{A + C'G + G'C + \Gamma + G'BG\}\underline{\delta} + o_p(N^{-1}), \tag{4b}$$

and

$$\Delta(\underline{q},\Phi(\underline{p})) = \Delta(\pi + N^{-1/2}\underline{\delta}_1,\Phi(\pi + N^{-1/2}\underline{\delta}_2)) = \Delta(\pi,\Phi(\pi)) + N^{-1/2}(\lambda'\underline{\delta}_1 + \eta'G\underline{\delta}_2) +$$
$$+ \ 1/2 \ N^{-1}\{\underline{\delta}_1'A\underline{\delta}_1 + 2 \ \underline{\delta}_1'CG\underline{\delta}_2 + \underline{\delta}_2'(\Gamma + G'BG)\underline{\delta}_2\} + o_p(N^{-1}). \tag{4c}$$

It follows that

$$N\{ \ E\{ \ \Delta(\pi,\Phi(\underline{p})) - \Delta(\pi,\Phi(\pi))\}\} = 1/2 \ \mathrm{tr} \ \{\Gamma + G'BG\}V + o(1), \tag{5a}$$

$$N\{E\{ \ \Delta(\underline{p},\Phi(\underline{p})) - \Delta(\pi,\Phi(\pi))\}\} = 1/2 \ \mathrm{tr} \ \{A + C'G + G'C + \Gamma + G'BG\}V + o(1), \tag{5b}$$

$$N\{E\{ \ \Delta(\underline{q},\Phi(\underline{p})) - \Delta(\pi,\Phi(\pi))\}\} = 1/2 \ \mathrm{tr} \ \{A + \Gamma + G'BG\}V + o(1). \tag{5c}$$

In deriving these expressions we have not used the fact that Φ is the projection of p in the metric Δ on Ω. The formulae (4) and (5) are true for any function of the proportions (which has sufficiently many derivatives). If we use the fact that Φ is a projection, then we can derive a formula for the matrix of partial derivatives G, using the implicit function theorem (compare Wolfe, 1976, or Abatzoglou, 1979, for closely related results). First introduce a local coordinate system $\Psi(\theta)$, with $\theta \ \varepsilon \ \Re^r$, in the point $\Phi(\pi)$. The matrix H contains the partial derivatives of Ψ with respect to the r coordinates, i.e. the columns of H are a basis for the tangent space of Ω in $\Phi(\pi)$. Also define the matrix ∇ with elements

$$\alpha_{st} = \Sigma_{k=1}^m \ \eta_k \ \partial^2\psi_k/\partial\theta_s\partial\theta_t, \tag{6}$$

where the second order derivatives are evaluated at $\Phi(\pi)$. Now the partials are given by

$$G = -H\{H'BH + \nabla\}^{-1}H'C. \tag{7}$$

Formula (7) can be substituted in (4) and (5), but we do not present the resulting expressions. We do point out that in the extreme cases S^{m-1} and π_0 we have G = I or G = 0. In both cases the map Φ is linear, and Γ = 0. This obviously leads to dramatic simplifications in (4) and (5).

In the general case it is somewhat tedious to use (5), because it involves the complicated matrix Γ of second derivatives of Φ. But it follows from (5) that

$$N\{ \ E\{ \ \Delta(\pi,\Phi(\underline{p})) - E\{ \ \Delta(\underline{p},\Phi(\underline{p}))\}\} = -1/2 \ \mathrm{tr} \ \{A + C'G + G'C\}V + o(1), \tag{8a}$$

$$N\{E\{\ \Delta(\underline{q},\Phi(\underline{p}))\ \text{-}\ E\{\ \Delta(\underline{p},\Phi(\underline{p}))\}\} = \text{-}\ \text{tr}\ G'CV + o(1). \tag{8b}$$

Thus if we define $\vartheta(p) = \{N\Delta(p,\Phi(p)) - \text{tr}\ C'GV\}/N$, with C, G, and V evaluated at $(p,\Phi(p))$, then

$$E\{\Delta(\underline{q},\Phi(\underline{p}))\} = E\{\vartheta(\underline{p})\} + o(N^{-1}). \tag{9a}$$

In the same way, with $\wp(p) = \{N\Delta(p,\Phi(p)) - 1/2\ \text{tr}\ AV - \text{tr}\ C'GV\}/N$, we have

$$E\{\Delta(\pi,\Phi(\underline{p}))\} = E\{\wp(\underline{p})\} + o(N^{-1}). \tag{9b}$$

This means that bias and distortion can be estimated consistently by using $\wp(\underline{p})$ and $\vartheta(\underline{p})$, for which we do not need the second order derivatives of Φ.

Best asymptotically normal theory

Now suppose the model is true. Then $\Phi(\pi) = \pi$, and expansions (4) simplify to

$$\Delta(\pi,\Phi(\underline{p})) = 1/2\ N^{-1}\underline{\delta}'G'BG\underline{\delta} + o_p(N^{-1}), \tag{10a}$$

$$\Delta(\underline{p},\Phi(\underline{p})) = 1/2\ N^{-1}\underline{\delta}'(I - G)'B(I - G)\underline{\delta} + o_p(N^{-1}), \tag{10b}$$

$$\Delta(\underline{q},\Phi(\underline{p})) = 1/2\ N^{-1}(\underline{\delta}_1 - G\underline{\delta}_2)'B(\underline{\delta}_1 - G\underline{\delta}_2) + o_p(N^{-1}). \tag{10c}$$

Thus $N\Delta(\pi,\Phi(\underline{p}))$, $N\ \Delta(\underline{p},\Phi(\underline{p}))$, and $N\ \Delta(\underline{q},\Phi(\underline{p}))$ are asymptotically quadratic forms in normal variables. We can compute their moments, and in particular

$$NE\{\Delta(\pi,\Phi(\underline{p}))\} = 1/2\ \text{tr}\ G'BGV + o(1), \tag{11a}$$

$$NE\{\ \Delta(\underline{p},\Phi(\underline{p}))\} = 1/2\ \text{tr}\ (I - G)'B(I - G)V + o(1), \tag{11b}$$

$$NE\{\ \Delta(\underline{q},\Phi(\underline{p}))\} = 1/2\ \text{tr}\ (V + G'VG)B + o(1). \tag{11c}$$

We now study the concept of *optimality* a bit more in detail, using bias and expected prediction error as criteria. The expected values in (11) still depend on B, which is a property of the separator Δ, and on G, which is a property of the estimator Φ. We know that $\Phi(p) = p$ for all $p \in \Omega$. Using a local coordinate system Ψ around π, it follows that $\Phi(\Psi(\theta)) = \Psi(\theta)$, with $\Psi(\theta_0) = \pi$. By differentiating this we see that $GH = H$, where $H = D\Psi(\theta_0)$, i.e. H is again a basis for the tangent space of Ω at π. But $GH = H$ can also be written as $G = H(H'\Pi^{-1}H)^{-1}H'\Pi^{-1} + TH'_\perp$, with $H_\perp$ a basis for the null space of H, i.e. the normal space to Ω at π. Now tr $BGVG' = $ tr $BH(H'\Pi^{-1}H)^{-1}H' + $ tr $BTH'_\perp VH_\perp T' \geq $ tr $BH(H'\Pi^{-1}H)^{-1}H'$. Here we have used the fact that B is positive definite, while $V\Pi^{-1}H = H$. Estimators Φ for which we have that $G = H(H'\Pi^{-1}H)^{-1}H'\Pi^{-1}$ are called BAN, or *best asymptotically normal*. Compare Neyman (1949), Wijsman (1959), Berkson (1980), Bemis and Bhapkar (1983), and LeCam (1986, section 11.10) for various aspects of the general theory of BAN estimation. The estimates are asymptotically normal, and they are are best in the class of F-consistent estimators, in the sense that their variance $(H'\Pi^{-1}H)^{-1}$ is minimal. If we apply this to (11) we find that

$$NE\{\Delta(\pi,\Phi(\underline{p}))\} = 1/2 \text{ tr } (H'\Pi^{-1}H)^{-1}H'BH + o(1), \tag{12a}$$

$$NE\{\Delta(\underline{p},\Phi(\underline{p}))\} = 1/2 \text{ tr } VB - 1/2 \text{ tr } (H'\Pi^{-1}H)^{-1}H'BH + o(1), \tag{12b}$$

$$NE\{\Delta(\underline{q},\Phi(\underline{p}))\} = 1/2 \text{ tr } VB + 1/2 \text{ tr } (H'\Pi^{-1}H)^{-1}H'BH + o(1), \tag{12c}$$

if Φ is BAN, and this is optimal, no matter how we choose B (i.e. no matter what separator we use).

The interesting question remains which separators give BAN estimates when minimized (also compare Taylor, 1953). From (8) we have, if the model is true, $G = H\{H'BH\}^{-1}H'B$. We see that Φ is BAN, for any model, if $B = \kappa\Pi^{-1}$ for some $\kappa \neq 0$. In this case we say that Δ is an *optimal multinomial separator*. It is *normalized* if $\kappa = 2$. A number of these normalized optimal multinomial separators are given by Rao (1973, section 5d.2). Thus if we derive our F-consistent estimator by projection using a separator, then we must choose an optimal separator to get BAN estimates. If the model is true, then the combination of an arbitrary normalized optimal separator and an arbitrary BAN estimate gives a technique for which the asymptotic distributions of the three statistics N $\Delta(\pi,\Phi(\underline{p}))$, N $\Delta(\underline{p},\Phi(\underline{p}))$, and N $\Delta(\underline{q},\Phi(\underline{p}))$ are central chi squares, with the appropriate numbers of degrees of freedom. In addition

$$NE\{\Delta(\pi,\Phi(\underline{p}))\} = r + o(1), \tag{13a}$$

$$NE\{\Delta(\underline{p},\Phi(\underline{p}))\} = (m - r - 1) + o(1), \tag{13b}$$

$$NE\{\Delta(\underline{q},\Phi(\underline{p}))\} = (m + r - 1) + o(1). \tag{13c}$$

This implies that $\{N\Delta(\underline{p},\Phi(\underline{p})) + 2r\}/N$ is a consistent estimate of the distortion, which seems a very simple way to justify the AIC-criterion of Akaike (1977). Compare also Sakamoto, Ishiguro, and Kitagawa (1986). Also $\{N\Delta(\underline{p},\Phi(\underline{p})) - (m - 2r - 1)\}/N$ is a consistent estimate of the bias.

Model comparison if the models are true

Let us now briefly review some of the techniques that have been proposed in statistics to compare models. These comparisons have usually been in terms of significance tests, and they try to answer the question 'is the model Ω true ?'. We have emphasized from the start that the answer to this question is simply 'no'. One does not need statistics to answer it, the answer follows from general considerations about the role of models. Our suggestion is to try and answer the question 'is the model Ω helpful in improving our predictions and descriptions ?'. This is a different question, and it is consequently not surprising that it may have a different answer. But a review of the classical procedures from our more general point of view is still useful.

Testing if a single model Ω is satisfactory can be interpreted as a problem of model choice: it compares Ω and S^{m-1}. The classical statistical technique for comparing Ω and S^{m-1} is to use the statistic $N\Delta(\underline{p},\Phi(\underline{p}))$, with Φ a BAN-estimator, and with Δ an optimal normalized multinomial separator. We already know that $N\Delta(\underline{p},\Phi(\underline{p}))$ is asymptotically a chi square with m - r - 1 degrees of freedom. The classical statistical procedure, due for r = 0 to Pearson and for general r to Fisher, tells us to reject the model Ω if $N\Delta(\underline{p},\Phi(\underline{p}))$ is too large, because this implies that 'either the model is untrue, or a very improbable event has occurred'. There is no reason to worry if you have trouble understanding the syllogism in the previous sentence. Nobody understands it. But the procedure can be interpreted quite simply in our terms. We evaluate a model by computing the integral from $N\Delta(\underline{p},\Phi(\underline{p}))$ to $+\infty$ of the χ^2_{m-r-1} distribution. The resulting 'P-value' must be as large as possible. It is difficult for most models to compete in P-value with the saturated model, because this has a P-value of one.

We now illustrate the procedure on our twin example. As the separator we take the one associated with the method of maximum likelihood. This is

$$\Delta(p,q) = 2 \Sigma_{j=1}^{m} \{p_j \ln p_j - p_j \ln q_j\}. \tag{14}$$

Thus $\lambda_j(p,q) = 2(1 + \ln p_j - \ln q_j)$ and $\eta_j(p,q) = -2p_j/q_j$. Moreover $A(p,q) = 2P^{-1}$, $C(p,q) = -2Q^{-1}$, and $B(p,q) = 2PQ^{-2}$. Here $P = \text{diag}(p)$ and $Q = \text{diag}(q)$. If $p = q$ then $\lambda(p,p) = 2u$, $\eta(p,p) = -2u$, where u has all elements equal to +1, and $A(p,p) = B(p,p) = -C(p,p) = 2P^{-1}$, which shows that the maximum likelihood estimator is BAN.

We start with the simple binomial model, which says that the probability of a boy is equal to the probability of a girl. The projection distance is $2N\{\underline{p} \ln \underline{p} + (1 - \underline{p}) \ln (1 - \underline{p}) + \ln 2\}$, which has a chi square distribution with one degree of freedom if the model is true. Of course for the saturated binomial model, which does not restrict the probability of a boy or girl, the projection distance is zero. Table 2 lists the projection distances, together with the P-values, for the six centres. According to the usual criteria we would accept the restrictive model for all centres, except perhaps for Zagreb. If we use the distortion as a criterion (assuming that the model is true) we see from (13) that the restrictive model is better than the saturated model if the projection distance is less than two. Again this is the case for all centres, except

	chi	P-value
Melbourne	1.00	.317
Sao Paolo	0.61	.438
Santiago	1.00	.317
Alexandria	0.02	.888
Hong Kong	0.02	.903
Zagreb	3.52	.061

Table 2: binomial model tests

Zagreb. This means that in all centres we estimate the probability of a boy to be equal to .5, but in Zagreb we estimate it to be $70/134 = .5224$. We see the damping effect of our use of models, which is basically the same as the shrinkage in empirical Bayes procedures, in ridge regression, or in Morris-Stein estimation.

Now we analyze the five models A-E for twin-pairs in the classical way. Consider model A, which is a one-dimensional quadratic manifold. The parametrisation in (1) is valid for all $0 \leq \omega \leq 1$. The maximum likelihood estimator of ω is, after some easy computation, given by $p_1 + p_3/2$. Thus $\phi_1(p) = (p_1 + p_3/2)^2$, $\phi_2(p) = (p_2 + p_3/2)^2$, and $\phi_3(p) = 2(p_1 + p_3/2)(p_2 + p_3/2)$. For model D, which is a line segment, the maximum likelihood estimator of λ is $p_1 + p_2 - p_3$. Thus $\phi_1(p) = \phi_2(p) = (1 + p_1 + p_2 - p_3)/4$, and $\phi_3(p) = (1 - p_1 - p_2 + p_3)/2$. For model E the only F-consistent estimator is $\phi_1(p) = \phi_2(p) = .25$ and $\phi_3(p) = .50$. Models B and C are somewhat problematical from the classical viewpoint. Model B says that there are no pairs of twins with different gender in the population. Thus they also cannot occur in any subset, and if only one occurs in the sample we reject B. No statistics is needed, only logic. Model C gives a maximum likelihood estimator equal to p if p is in the convex region defining the model, and equal to the estimate under model A if p is outside the region. For any π interior to the model the probability that p is interior tends to one, and thus the projection distance tends to zero with probability one. If π is on the model A boundary, then the projection distance has the mixture-distribution $\text{prob}(p \notin C)\chi_1^2 + \text{prob}(p \in C)\chi_0^2 = \text{prob}(p \notin C)\chi_1^2$.

In Table 3a we have listed N times the projection distances for the models A, D, and E. The corresponding quantitites for model B are all 'infinite', those of model C are all zero. According to classical statistical theory the projection distances have chi square distributions with one (models A and D) or two (model E) degrees of freedom. Table 3b uses these asymptotic distributions to convert the chi square values to probabilities. This puts them on a convenient scale, and makes them comparable. The estimate of the distance between the observed and the expected value under the (true) model is thus corrected for the dimensionality of the model. Table 3 shows clearly that according to the classical analysis model D is the best one. It also shows that on the probability scale model E is better than A. One can interpret this according to the classical, although mysterious, syllogism quoted above. One can also think in terms of distance estimates between models and data, transformed to a convenient scale. But an interpretation of this scale in our replication framework is possible only if the model is true.

	model A	model D	model E
Melbourne	10.40	0.76	11.40
Sao Paulo	11.37	0.49	11.98
Santiago	33.39	0.73	34.39
Alexandria	12.24	0.02	12.26
Hong Kong	26.97	0.01	26.99
Zagreb	5.24	2.79	8.77

Table 3a: chi squares

	model A	model D	model E
Melbourne	.001	.383	.003
Sao Paulo	.001	.484	.003
Santiago	.001	.393	.001
Alexandria	:001	.888	.002
Hong Kong	.001	.920	.001
Zagreb	.022	.095	.012

Table 3b: probability transform

Another comparison is based on Akaike's AIC. As we have seen this amounts to adding two to the chi squares of models A and D, and zero to those of model E. The resulting quantity estimates the distortion, but again only if the model is true. Clearly this does not change the conclusions a great deal. Model D is still the best, by far, and on the AIC scale E is also slightly better than A, except for Zagreb.

Model evaluation using separator statistics

We now continue with our evaluation of models, but it is no longer assumed that the model, or one of the models, is true. Within classical statistics there has been at least one major development which applies in this more general situation. This the theory of *Pitman powers*. In this theory one does not assume that $\pi \in \Omega$. The assumption is that there is a sequence Ω_N of models, giving rise to a sequence Φ_N of projections. We assume that the approximation errors $\Delta(\pi,\Phi_N(\pi))$ tend to zero in such a way that $N\Delta(\pi,\Phi_N(\pi))$ tends to a constant Δ_0. More precisely the assumptions guarantee that $N\Delta(\underline{p},\Phi_N(\underline{p}))$ converges in law to a noncentral chi square with $m - r - 1$ degrees of freedom, and noncentrality parameter Δ_0. Thus $NE\{\Delta(\underline{p},\Phi_N(\underline{p}))\} = (m - r - 1) + \Delta_0 + o(1)$. Verbeek (1984) has suggested to use the *noncentrality*, given by $\{N\Delta(\underline{p},\Phi_N(\underline{p})) - (m - r - 1)\}/N$, as an index of model fit. It is convenient to think of the noncentrality as an estimate of the approximation error $\Delta(\pi,\Phi_N(\pi))$ for models which are not too false. Nevertheless the use of Pitman powers in actual practical work seems a bit difficult. Who actually works with a sequence of models ?

A more direct approach is to go back to the formulas (5) to (9). In the case of the maximum likelihood separator (14) we find that $1/2 \, \text{tr} \, AV = (m - 1)$ and $\text{tr} \, C'GV = -2 \, \text{tr} \, H'BH(H'BH + \nabla)^{-1}$. Thus we can estimate the bias by

$$\wp(\underline{p}) = \{N\Delta(\underline{p},\Phi(\underline{p})) + 2 \, \text{tr} \, (H'BH + \nabla)^{-1}H'BH \, - (m - 1)\}/N, \tag{15a}$$

and the distortion by

$$\vartheta(\underline{p}) = \{N\Delta(\underline{p},\Phi(\underline{p})) + 2 \, \text{tr} \, (H'BH + \nabla)^{-1}H'BH\}/N. \tag{15b}$$

In (15) $B = 2PQ^{-2}$ is evaluated in (p,q) equal to $(\underline{p},\Phi(\underline{p}))$. The basis H is evaluated in $\underline{\theta}$, where $\Psi(\underline{\theta}) = \Phi(\underline{p})$. Matrix ∇ is evaluated at the same point. If $\nabla = 0$, which happens for instance if Ω is an affine manifold, then $\text{tr} \, (H'BH + \nabla)^{-1}H'BH = r$, and (15) simplifies accordingly. This gives a justification for using the AIC, even if the model is not true.

We can use formula (5) to estimate the approximation error. Using the simplifications resulting from the use of the maximum likelihood method we find

$$\aleph(\underline{p}) = \{N\Delta(\underline{p},\Phi(\underline{p})) - (m - 1) + 2 \, \text{tr} \, (H'BH + \nabla)^{-1}H'BH -$$
$$- \text{tr} \, (H'BH + \nabla)^{-1}H'BH(H'BH + \nabla)^{-1}H'BH \, - 1/2 \, \text{tr} \, \Gamma V\}/N \tag{16}$$

This satisfies $E(\aleph(\underline{p})) = \Delta(\pi,\Phi(\pi)) + o(N^{-1})$. It is related, but not identical, to the noncentrality proposed by Verbeek, and it becomes identical to it in the case that ∇ and Γ are zero.

Let us now apply formulas (15) and (16) to the twin example. For the binomial model (and for saturated and zero-dimensional models in general) we find that indeed ∇ and Γ are zero. Thus $\wp(\underline{p}) = \{N\Delta(\underline{p},\Phi(\underline{p})) \, - (m - 2r - 1)\}/N$, and $\vartheta(\underline{p}) = \{N\Delta(\underline{p},\Phi(\underline{p})) + 2r\}/N$. Moreover $\aleph(\underline{p}) = \{N\Delta(\underline{p},\Phi(\underline{p})) - (m - r - 1)\}/N$. For models D and E we also have ∇ and Γ equal to zero, and the same formulas apply. For model A the situation is more complicated. Table 4a has the relevant information, but the computations needed to arrive at the numbers in the table, especially those for estimating the approximation error, are not simple. They will be even more complicated for models with many more cells than the three in our example. Table 4a shows that the bias and distortion are larger for model A than for the saturated model. In Table 4b we show the comparable statistics for (zero-dimensional) model E. Clearly E is preferable to A, even in terms of estimated approximation error (although of course the true approximation error for A is lower).

	bias	prediction	approximation
Melbourne	.113	.133	.099
Sao Paulo	.056	.066	.050
Santiago	.142	.150	.136
Alexandria	.032	.037	.029
Hong Kong	.223	.239	.211
Zagreb	.070	.094	.055

Table 4a: separator statistics model A.

	bias	prediction	approximation
Melbourne	.086	.106	.086
Sao Paulo	.044	.054	.044
Santiago	.130	.139	.130
Alexandria	.026	.031	.026
Hong Kong	.200	.216	.200
Zagreb	.040	.064	.040

Table 4b: separator statistics model E.

Use of Jackknife-type methods

The separator statistics used in the previous section are fairly difficult to compute, even if we use the simplifications that result from using the maximum likelihood method. For complicated models evaluating the second derivatives may be a painful process. A possible way out is to use resampling methods such as the Bootstrap and the Jackknife. Compare Efron (1982) for a nice review of these methods. In this paper we do not go into the philosophy of resampling, we merely use the methods as computational tools that use finite difference approximations to the derivatives, and that consequently can be used to approximate the separator statistics. We also restrict our attention to Jackknife (i.e. leave-one-out) methods. These are computationally far less demanding than the Bootstrap methods, and in the cases in which we have compared the two they give virtually identical results.

Define $q_{[j]} = p + (N-1)^{-1}(p - e_j)$, with e_j the jth unit vector. Then

$$\Delta(q_{[j]}, \Phi(p)) = \Delta(p, \Phi(p)) + (N-1)^{-1}\lambda'(p - e_j) + 1/2 \ (N-1)^{-2} \ (p - e_j)'A(p - e_j) + o((N-1)^{-2}), \qquad (17)$$

and thus

$$\Sigma_{j=1}^m \, p_j \Delta(q_{[j]}, \Phi(p)) = \Delta(p, \Phi(p)) + 1/2 \, (N-1)^{-2} \, \text{tr} \, AV + o((N-1)^{-2}), \tag{18}$$

or

$$(N-1)^2 \{ \Sigma_{j=1}^m \, p_j \Delta(q_{[j]}, \Phi(p)) - \Delta(p, \Phi(p)) \} = 1/2 \, \text{tr} \, AV + o(1). \tag{19a}$$

In the same way

$$(N-1)^2 \{ \Sigma_{j=1}^m \, p_j \Delta(p, \Phi(q_{[j]})) - \Delta(p, \Phi(p)) \} = 1/2 \, \text{tr} \, (\Gamma + G'BG)V + o(1), \tag{19b}$$

and

$$(N-1)^2 \{ \Sigma_{j=1}^m \, p_j \Delta(q_{[j]}, \Phi(q_{[j]})) - \Delta(p, \Phi(p)) \} =$$

$$= 1/2 \, \text{tr} \, (A + \Gamma + G'BG + G'C + C'G)V + o(1). \tag{19c}$$

It is clear that, under suitable regularity assumptions, we can use the quantities on the left of (19) to estimate the quantities on the right. If we combine (19) with (5), we see that estimating the approximation error, the bias and the distortion becomes quite simple. We first show this for the approximation error. Combining (5b) and (19c) shows that

$$E\{N\Delta(p,\Phi(p)) - (N-1)\Sigma_{j=1}^m \, p_j \Delta(q_{[j]}, \Phi(q_{[j]}))\} = \Delta(\pi, \Phi(\pi)) + o((N-1)^{-1}). \tag{20}$$

This is the classical multinomial Jackknife result, but it is remarkable (and very satisfactory) that it can be generalized easily to deal with bias and distortion. Indeed, from (19b) and (19c),

$$(N-1)^2 \{ \Sigma_{j=1}^m \, p_j \Delta(p, \Phi(q_{[j]})) - \Sigma_{j=1}^m \, p_j \Delta(q_{[j]}, \Phi(q_{[j]})) \} = -1/2 \, \text{tr} \, (A + G'C + C'G)V + o(1), \tag{21}$$

and thus

$$E\{\Delta(p,\Phi(p)) + (N-1)\{\Sigma_{j=1}^m \, p_j \Delta(p,\Phi(q_{[j]})) - \Sigma_{j=1}^m \, p_j \Delta(q_{[j]}, \Phi(q_{[j]}))\}\} =$$

$$= E\{\Delta(\pi, \Phi(p))\} + o((N-1)^{-1}). \tag{22}$$

In the same way, from (19a) and (21),

$$(N-1)^2 \{ \Sigma_{j=1}^m \, p_j \Delta(q_{[j]}, \Phi(p)) + \Sigma_{j=1}^m \, p_j \Delta(p, \Phi(q_{[j]})) - \Delta(p, \Phi(p)) - \Sigma_{j=1}^m \, p_j \Delta(q_{[j]}, \Phi(q_{[j]})) \}$$

$$= - \text{tr } G'CV + o(1), \tag{23}$$

and thus

$$E\{N\Delta(\underline{p},\Phi(\underline{p})) + (N-1)\{\Sigma_{j=1}^m p_j\Delta(q_{[j]},\Phi(p)) + \Sigma_{j=1}^m p_j\Delta(p,\Phi(q_{[j]})) - \Sigma_{j=1}^m p_j\Delta(q_{[j]},\Phi(q_{[j]}))\}\} =$$

$$= E\{\Delta(\underline{q},\Phi(\underline{p}))\} + o((N-1)^{-1}). \tag{24}$$

These formulae may look complicated, but they are actually quite easy to use. The computational burden is that we must evaluate our estimate Φ not once, but $m + 1$ times, with m the number of cells in the multinomial experiment. This could be a large amount of work, especially in very large multiway tables, but in such very large tables using asymptotic theory is of doubtful value anyway. Observe that these formulas are perfectly general, in the sense that they use no properties of the likelihood separator and of the maximum likelihood estimate. They also do not assume, of course, that the model is true.

We shall illustrate our formulae by applying (20) to our example, in particular to models A, D, and E. Table 5 lists the relevant results. Table 5a actually presents the Jackknife result derived from (20), and

	model A	model D	model E
Melbourne	.099	-.003	.096
Sao Paulo	.050	-.002	.047
Santiago	.136	-.001	.134
Alexandria	.029	-.003	.026
Hong Kong	.211	-.008	.192
Zagreb	.055	.022	.082

Table 5a: Estimated approximation errors
using the Jackknife

	model A	model D	model E
Melbourne	.099	-.003	.096
Sao Paulo	.050	-.002	.047
Santiago	.136	-.001	.134
Alexandria	.029	-.002	.026
Hong Kong	.211	-.008	.200
Zagreb	.055	.022	.082

Table 5b: Estimated approximation errors
using the Bootstrap

Table 5b gives the corresponding Bootstrap estimates (from results not presented here). It is clear that the results are very close indeed, and because the Bootstrap is computationally so much more demanding it does not seem to be a wise choice in this context. We can also compare the results for models A and E in Table 5 with the last columns of Tables 4a and 4b. The correspondence is very satisfactory.

Conclusion

It is clear from our discussion above that the expansions we have derived are valuable tools to answer the simple question whether the model is useful or not. But the meaning of 'useful' must be specified in at last two ways before we can actually answer the question. In the first place a model can be useful for bias reduction and for prediction. As we have seen above these two forms of usefulness tend to be contradictory, in the sense that 'large' models are good for bias reduction and bad for prediction, while for 'small' models it is the other way around. The solution of classical statistics is to choose from the class of true models only. This approach does not make sense to us, although it is possible to interpret the classical procedures in our somewhat wider framework. But in this wider framework the classical procedures may not give good estimates of bias and expected prediction error.

A second specification we have to choose before we can answer the question about the usefulness of a model is the choice of a separator. The one suggested by the method of maximum likelihood is convenient, the class of methods suggested by BAN-theory is also attractive, but essentially the choice is open. Small bias in terms of one separator does not necessarily imply small bias in terms of another one. It is quite possible that results can be derived, in our general framework, which show that some separators are asymptotically or uniformly preferable to others (think of second order efficiency), but we have no results in this direction.

The question if a model is useful or not can be coupled with the question of computing a good estimate of the unknown probabilities in the multinomial model. If we decide that a particular model is preferable to the saturated model, then we can replace the sample proportion by the estimates under the model restrictions. We have seen, in our simple example, that considering a number of models works in the same way as a prior distribution, and produces shrinkage estimators very much like the empirical Bayes estimators familiar from other statistical work. No explicit prior distribution is considered, but a discrete class of models must be chosen for consideration, and of course this choice is to some extent 'subjective'. It is a useful subject of study to find out if these shrunken estimators are in some sense better than the usual estimators, although the method with which they are constructed already seems to answer this question in the affirmative.

We think that the primary contribution of this paper is, or should be, that it tries to make people more careful about assuming models to be true (whatever that means), about using standard statistical reasoning (as in hypothesis testing), and about using probability models without clearly specifying the framework of replication. Whether specific models are actually false or not is not interesting. All models are false. The question is how false, and whether their being false actually implies that they are not useful. And whether the resulting statements are consequently not interesting. The independence assumption, for example, which is at the basis of most work in statistics, cannot really be falsified. As we have seen, the independence assumption merely corresponds with a particular framework of replication, for which we have to decide whether it is relevant or not. We make statements about all samples with replacement from the population of the Netherlands, for instance. If such statements are really relevant for the particular policy issue we are studying is something which certainly must be decided with a great deal of care. For the more complicated frameworks, such as those with continuous variables and non-identically distributed observations, it may be most difficult to convince a sceptical user that our statements are indeed relevant for his problem.

Another more specific contribution of the paper is that it shows that looking at the projection distance is only a first crude apprximation. The interesting statistics, bias and distortion, can easily be approximated more precisely, both by using additional terms in the expansions and by using resampling methods. In particular we have argued that the distortion, which is the expected value of the prediction error, is at least as interesting as the bias, the expected value of the estimation error. In general the two statistics lead to a different ordering of models, and consequently also to different estimates of π. We have tried to justify the AIC statistic, used to estimate distortion, and the deviance statistic, used to estimate approximation error used by Verbeek, without assuming the model to be even approximately true.

References

Abatzoglou,T, The Metric Projection on C^2 Manifolds in Banach Spaces. **Journal of Approximation Theory**, 26, 1979, 204-211.

Akaike, H., On Entropy Maximization Principle. In P.R. Krishnaiah (ed.), **Applications of Statistics**, Amsterdam, North Holland Publishing Company, 1977

Andersen. E.B., **Discrete Statistical Models with Social Science Applications.** Amsterdam, North Holland Publishing Co, 1980.

Bemis, K.G. & Bhapkar, V.P., On BAN Estimators for Chi Squared Test Criteria, **The Annals of Statistics**, 11, 1983, 183-196.

Benzécri, J.P. et al., **L'Analyses des Données.** Paris, Dunod, 1973.

Berkson, J., Minimum Chi-Square, not Maximum Likelihood (with Discussion). **The Annals of Statistics**, 8, 1980, 457-487.

Box, G.E.P., Some problems of Statistics and Everyday Life. **Journal of the American Statistical Association**, 74, 1979, 1 4.

De Leeuw, J., Models for Data. **Kwantitatieve Methoden**, 5, 1984, 17-30.

Efron, B. **The Bootstrap, the Jackknife, and other Resampling Plans.** Philadelphia, SIAM, 1982.

Gifi, A., **Nonlinear Multivariate Analysis.** Leiden, Department of Data Theory FSW/RUL, 1981

Guttman, L., The Illogic of Statistical Inference for Cumulative Science. **Applied Stochastic Models and Data Analysis**, 1, 1985, 3-10.

Hemelrijk, J., Back to the Laplace definition. **Statistica Neerlandica**, 22, 1968, 13-21.

Kalman, R.E., Identifiability and Modeling in Econometrics. In P.R. Krishnaiah (ed.), **Developemnts in Statistics**, Amsterdam, North Holland Publishing Company, 1983.

Le Cam, L., **Asymptotic Methods in Statistical Decision Theory.** Berlin, Springer, 1986.

McCullagh , P. & Nelder, J.A., **Generalized Linear Models.** London, Chapman and Hall, 1983.

Nelder, J.A., The Role of Models in Official Statistics. **Eurostat News**, 1984, special number on Recent Developments in the Analysis of Large-Scale Data Sets.

Neyman, J., Contributions to the Theory of the χ^2 Test. **Proceedings Berkeley Symposium**, 1, 1949, 239-273.

Rao, C.R., **Linear Statistical Inference and its Applications**, New York, Wiley, 1973.

Sakamoto, Y., Ishiguro, M., & Kitagawa, G., **Akaike Information Criterion Statistics**. Dordrecht, Reidel, 1986.

Taylor, W.F., Distance Functions and Regular Best Asymptotically Normal Estimates. **Annals of Mathematical Statistics**, 24, 1953, 85-92.

Tukey, J.W., We need both Exploratory and Confirmatory. **American Statistician**, 34, 1980, 23-25.

Verbeek, A. The Geometry of Model Selection in Regression. In T.K. Dijkstra (ed.), **Misspecification Analysis**, Berlin, Springer, 1984.

Willems, J.C., **From Time Series to Linear System**, Mathematical Institute, University of Groningen, 1986

Wolfe, J.M., Differentiability of Nonlinear Best Approximation Operators in a Real Inner Product Space. **Journal of Approximation Theory**, 16, 1976, 341-346.

Wijsman, R.A., On the Theory of B.A.N. estimates. **Annals of Mathematical Statistics**, 30, 1959, 185-191. Correction idem, 1268-1270.

Vol. 264: Models of Economic Dynamics. Proceedings, 1983. Edited by H.F. Sonnenschein. VII, 212 pages. 1986.

Vol. 265: Dynamic Games and Applications in Economics. Edited by T. Başar. IX, 288 pages. 1986.

Vol. 266: Multi-Stage Production Planning and Inventory Control. Edited by S. Axsäter, Ch. Schneeweiss and E. Silver. V, 264 pages. 1986.

Vol. 267: R. Bemelmans, The Capacity Aspect of Inventories. IX, 165 pages. 1986.

Vol. 268: V. Firchau, Information Evaluation in Capital Markets. VII, 103 pages. 1986.

Vol. 269: A. Borglin, H. Keiding, Optimality in Infinite Horizon Economies. VI, 180 pages. 1986.

Vol. 270: Technological Change, Employment and Spatial Dynamics. Proceedings 1985. Edited by P. Nijkamp. VII, 466 pages. 1986.

Vol. 271: C. Hildreth, The Cowles Commission in Chicago, 1939–1955. V, 176 pages. 1986.

Vol. 272: G. Clemenz, Credit Markets with Asymmetric Information. VIII, 212 pages. 1986.

Vol. 273: Large-Scale Modelling and Interactive Decision Analysis. Proceedings, 1985. Edited by G. Fandel, M. Grauer, A. Kurzhanski and A.P. Wierzbicki. VII, 363 pages. 1986.

Vol. 274: W.K. Klein Haneveld, Duality in Stochastic Linear and Dynamic Programming. VII, 295 pages. 1986.

Vol. 275: Competition, Instability, and Nonlinear Cycles. Proceedings, 1985. Edited by W. Semmler. XII, 340 pages. 1986.

Vol. 276: M.R. Baye, D.A. Black, Consumer Behavior, Cost of Living Measures, and the Income Tax. VII, 119 pages. 1986.

Vol. 277: Studies in Austrian Capital Theory, Investment and Time. Edited by M. Faber. VI, 317 pages. 1986.

Vol. 278: W.E. Diewert, The Measurement of the Economic Benefits of Infrastructure Services. V, 202 pages. 1986.

Vol. 279: H.-J. Büttler, G. Frei and B. Schips, Estimation of Disequilibrium Models. VI, 114 pages. 1986.

Vol. 280: H.T. Lau, Combinatorial Heuristic Algorithms with FORTRAN. VII, 126 pages. 1986.

Vol. 281: Ch.-L. Hwang, M.-J. Lin, Group Decision Making under Multiple Criteria. XI, 400 pages. 1987.

Vol. 282: K. Schittkowski, More Test Examples for Nonlinear Programming Codes. V, 261 pages. 1987.

Vol. 283: G. Gabisch, H.-W. Lorenz, Business Cycle Theory. VII, 229 pages. 1987.

Vol. 284: H. Lütkepohl, Forecasting Aggregated Vector ARMA Processes. X, 323 pages. 1987.

Vol. 285: Toward Interactive and Intelligent Decision Support Systems. Volume 1. Proceedings, 1986. Edited by Y. Sawaragi, K. Inoue and H. Nakayama. XII, 445 pages. 1987.

Vol. 286: Toward Interactive and Intelligent Decision Support Systems. Volume 2. Proceedings, 1986. Edited by Y. Sawaragi, K. Inoue and H. Nakayama. XII, 450 pages. 1987.

Vol. 287: Dynamical Systems. Proceedings, 1985. Edited by A.B. Kurzhanski and K. Sigmund. VI, 215 pages. 1987.

Vol. 288: G.D. Rudebusch, The Estimation of Macroeconomic Disequilibrium Models with Regime Classification Information. VII, 128 pages. 1987.

Vol. 289: B.R. Meijboom, Planning in Decentralized Firms. X, 168 pages. 1987.

Vol. 290: D.A. Carlson, A. Haurie, Infinite Horizon Optimal Control. XI, 254 pages. 1987.

Vol. 291: N. Takahashi, Design of Adaptive Organizations. VI, 140 pages. 1987.

Vol. 292: I. Tchijov, L. Tomaszewicz (Eds.), Input-Output Modeling. Proceedings, 1985. VI, 195 pages. 1987.

Vol. 293: D. Batten, J. Casti, B. Johansson (Eds.), Economic Evolution and Structural Adjustment. Proceedings, 1985. VI, 382 pages. 1987.

Vol. 294: J. Jahn, W. Krabs (Eds.), Recent Advances and Historical Development of Vector Optimization. VII, 405 pages. 1987.

Vol. 295: H. Meister, The Purification Problem for Constrained Games with Incomplete Information. X, 127 pages. 1987.

Vol. 296: A. Börsch-Supan, Econometric Analysis of Discrete Choice. VIII, 211 pages. 1987.

Vol. 297: V. Fedorov, H. Läuter (Eds.), Model-Oriented Data Analysis. Proceedings, 1987. VI, 239 pages. 1988.

Vol. 298: S.H. Chew, Q. Zheng, Integral Global Optimization. VII, 179 pages. 1988.

Vol. 299: K. Marti, Descent Directions and Efficient Solutions in Discretely Distributed Stochastic Programs. XIV, 178 pages. 1988.

Vol. 300: U. Derigs, Programming in Networks and Graphs. XI, 315 pages. 1988.

Vol. 301: J. Kacprzyk, M. Roubens (Eds.), Non-Conventional Preference Relations in Decision Making. VII, 155 pages. 1988.

Vol. 302: H.A. Eiselt, G. Pederzoli (Eds.), Advances in Optimization and Control. Proceedings, 1986. VIII, 372 pages. 1988.

Vol. 303: F.X. Diebold, Empirical Modeling of Exchange Rate Dynamics. VII, 143 pages. 1988.

Vol. 304: A. Kurzhanski, K. Neumann, D. Pallaschke (Eds.), Optimization, Parallel Processing and Applications. Proceedings, 1987. VI, 292 pages. 1988.

Vol. 305: G.-J.C.Th. van Schijndel, Dynamic Firm and Investor Behaviour under Progressive Personal Taxation. X, 215 pages. 1988.

Vol. 306: Ch. Klein, A Static Microeconomic Model of Pure Competition. VIII, 139 pages. 1988.

Vol. 307: T.K. Dijkstra (Ed.), On Model Uncertainty and its Statistical Implications. VII, 138 pages. 1988.

T. Vasko (Ed.)

The Long-Wave Debate

Selected papers from an IIASA (International Institute for Applied
Systems Analysis) International Meeting on Long-Term Fluctuations in
Economic Growth: Their Causes and Consequences, Held in Weimar,
German Democratic Republic, June 10–14, 1985

1987. 128 figures. XVII, 431 pages. ISBN 3-540-18164-4

Contents: Concepts and Theories on the Interpretation of Long-Term
Fluctuations in Economic Growth. – Technical Revolutions and Long
Waves. – The Role of Financial and Monetary Variables in the Long-
Wave Context. – Modeling the Long-Wave Context. – Modeling the
Long-Wave Phenomenon. – List of Participants.

I. Boyd, J. M. Blatt

Investment Confidence and Business Cycles

1988. 160 pages. ISBN 3-540-18516-X

Contents: Introduction and brief summary. – A brief historical survey of
the trade cycle. – Literature on confidence. – The dominant theories. –
A first look at the new model. – Confidence. – Description of the model. –
The longer run. – Some general remarks. – Appendices. – References. –
Index.

M. Faber, H. Niemes, G. Stephan

Entropy, Environment and Resources

An Essay in Physico-Economics

With the cooperation of L. Freytag

Translated from the German by I. Pellengahr

1987. 33 figures. Approx. 210 pages. ISBN 3-540-18248-9

The special features of the book are that the authors utilize a natural
scientific variable, entropy, to relate the economic system and the environment, that environmental protection and resource use are analyzed in
combination, and that a replacement of techniques over time is analyzed.
A novel aspect is that resource extraction is interpreted as a reversed
diffusion process. Thus a relationship between entropy, energy and resource concentration is established.

E. van Damme

Stability and Perfection of Nash Equilibria

1987. 105 figures. Approx. 370 pages. ISBN 3-540-17101-0

Contents: Introduction. – Games in Normal Form. – Matrix and Bimatrix
Games. – Control Costs. – Incomplete Information. – Extensive Form
Games. – Bargaining and Fair Division. – Repeated Games. – Evolu-
tionary Game Theory. – Strategic Stability and Applications. – References. – Survey Diagrams. – Index.

Springer-Verlag
Berlin Heidelberg New York
London Paris Tokyo